A Language School as a Complex System

Achilleas Kostoulas

A Language School as a Complex System

Complex Systems Theory in English Language Teaching

PETER LANG

Bibliographic Information published by the Deutsche Nationalbibliothek
The Deutsche Nationalbibliothek lists this publication in the Deutsche Nationalbibliografie; detailed bibliographic data is available in the internet at http://dnb.d-nb.de.

Library of Congress Cataloging-in-Publication Data
A CIP catalog record for this book has been applied for at the Library of Congress.

Cover image: © ADDICTIVE STOCK / Fotolia.com

ISBN 978-3-631-73568-8 (Print)
E-ISBN 978-3-631-73571-8 (E-PDF)
E-ISBN 978-3-631-73572-5 (EPUB)
E-ISBN 978-3-631-73573-2 (MOBI)
DOI 10.3726/b11892

© Peter Lang GmbH
Internationaler Verlag der Wissenschaften
Berlin 2018
All rights reserved.

Peter Lang – Berlin · Bern · Bruxelles · New York ·
Oxford · Warszawa · Wien

All parts of this publication are protected by copyright. Any utilisation outside the strict limits of the copyright law, without the permission of the publisher, is forbidden and liable to prosecution. This applies in particular to reproductions, translations, microfilming, and storage and processing in electronic retrieval systems.

This publication has been peer reviewed.

www.peterlang.com

Table of Contents

About the Author 9

List of Tables 11

List of Figures 13

List of Frequently Used Abbreviations 15

1 Motivations, background, and queries 19

   1.1. Complex systems in education 20
      1.1.1. What are complex systems? 20
      1.1.2 Describing complex systems 21

   1.2 Setting the scene 24
      1.2.1 ELT in state education 24
      1.2.2 The private ELT sector 28
      1.2.3 The school 30

   1.3 Outline of this book 32

2 About complexity 35

   2.1 Establishing a new discipline 36

   2.2 But what is complexity? 37

   2.3 Characteristics of complex systems 40
      2.3.1 Complex systems are made up of heterogeneous components 41
      2.3.2 Complex systems have ambiguous boundaries 42
      2.3.3 Complex systems shape each other 43
      2.3.4 Complex systems produce emergent phenomena 44
      2.3.5 Complex systems tend to be very resilient 45
      2.3.6 Complex systems connect to their past states in curious ways 46

   2.4 Applications of CST 47
      2.4.1 CST in the study of language and language acquisition 48
      2.4.2 CST in education and ELT 50
      2.4.3 CST in the psychology of language learning and teaching 51
      2.4.4 CST as connective tissue among disciplines 53

   2.5 Ways forward 55

3 Exploring the school's state space 61

   3.1 The linguistic dimension: understandings of language 62
         3.1.1 The Standard Language ideology 63
         3.1.2 World Englishes 64
         3.1.3 English as a Lingua Franca 67

   3.2. The pedagogical dimension: teaching and learning approaches 70
         3.2.1 The transmissive approach 71
         3.2.2 The communicative approach 74
         3.2.3 The post-method approach 77

   3.3 The political dimension: visions of society 79
         3.3.1 Neutrality: seeing no evil 80
         3.3.2 Awareness: developing a critical understanding 81
         3.3.3 Resistance: empowering teachers and learners 84

   3.4 Constraining structures 86

4 Tracing the affordance landscape 89

   4.1 Learning materials creating affordances 89

   4.2 Overview of the learning materials 92

   4.3 Affordances in the grammar activities 96

   4.4 Affordances in the vocabulary activities 99

   4.5 Affordances in the reading and writing activities 103

   4.6 Affordances in the listening and speaking activities 106

   4.7 Putting it all together 109

5 Driving activity in the system 113

   5.1 Defining intentionality 113
         5.1.1 Intentionality is collective 114
         5.1.2 Intentionality is nested 114
         5.1.3 Intentionality is emergent 115
         5.1.4 Intentionality is generative 116

   5.2 Proving proficiency 117
         5.2.1 What was so important about certification? 117

  5.2.2 What certification options were available? 121
  5.2.3 On hierarchies and determinism 126

 5.3 Finding a place to belong 127

 5.4 Teaching about England 130
  5.4.1 Anglophile attitudes 131
  5.4.2 Anglocentric input 133
  5.4.3 The importance of local context 134

 5.5 Being the best 136
  5.5.1 Limitations of the state education system 136
  5.5.2 A culture of accountability 138
  5.5.3 Introduction of TEYL 140
  5.5.4 Complex interactions, unpredictable outcomes 142

 5.6 Preventing change 143
  5.6.1 The 'Greek reality' 143
  5.6.2 Locally acquired knowledge 144
  5.6.3 Primacy of practice 146
  5.6.4 Incompatible beliefs working together 147

 5.7 Dynamics of intentions 148

6 The shape of teaching and learning 153

 6.1 About attractors 154
  6.1.1 Terminological ground-clearing 155
  6.1.2 The shape of teaching and learning 156

 6.2 Reading and Vocabulary 159
  6.2.1 Prompt 160
  6.2.2 Reading and Listening 162
  6.2.3 Reading aloud 162
  6.2.4 Reading comprehension 165
  6.2.5 Vocabulary consolidation 166
  6.2.6 Practice 168
  6.2.7 Dictation 168
  6.2.8 Emergence in teaching and learning 169

 6.3 Traditional Grammar 171
  6.3.1 Prompting 173
  6.3.2 Explanation 173
  6.3.3 Practice 175

    6.3.4 Application 179
    6.3.5 Emergence as a localised phenomenon 180

6.4 Process-based Writing 181
    6.4.1 Collaborative writing tasks 183
    6.4.2 Individual writing practice 186
    6.4.3 Feedback 187
    6.4.4 Reorienting the system 188

6.5 An evolving system 189

7 Using complexity to describe a language school 193

7.1 Situating a system in place and time 193
    7.1.1 Rethinking boundaries 193
    7.1.2 Thinking about timescales 195

7.2 Looking at the structure of the system 197
    7.2.1 Some tentative findings… 197
    7.2.2 …and some theoretical insights 199

7.3 Examining the intentionalities that drive the system 200
    7.3.1 Defining intentionality 201
    7.3.2 Insights associated with the study of intentionalities 202

7.4 Pedagogical activity emerging in the system 204
    7.4.1 Pedagogical activity as an outcome of intentionalities
          and resources 204
    7.4.2 The role of local context 205
    7.4.3 ELT and globalisation 206

7.5 Future directions 209

Appendix. Methodological remarks 211

References 219

Index 249

About the Author

Achilleas Kostoulas taught English in Greece before moving on to language teacher education. He completed a PhD at the University of Manchester. He teaches courses in ELT and Applied Linguistics at the University of Graz. His research interests focus on the psychology of language learning and teaching.

List of Tables

Tab. 1.1	ELT provision in primary education
Tab. 1.2	ELT provision in secondary education
Tab. 1.3	ELT provision in the private sector
Tab. 1.4	Overview of the General English programmes
Tab. 3.1	Constraining structures in the state space
Tab. 4.1	Structure of a typical module (Junior)
Tab. 4.2	Structure of a typical module (Senior – New)
Tab. 4.3	Structure of a typical module (Senior – Old)
Tab. 4.4	Structure of a typical module (Exam)
Tab. 4.5	Extracts of content pages
Tab. 4.6	Format of monolingual vocabulary lists
Tab. 4.7	Spatial variation of affordances
Tab. 4.8	Phase variation of affordances
Tab. 4.9	Diachronic variation of affordances
Tab. 6.1	Prevalence of attractors in the curriculum
Tab. 6.2	Overview of a Reading and Vocabulary sequence
Tab. 6.3	Format of vocabulary notebooks
Tab. 6.4	Overview of a Traditional Grammar sequence
Tab. 6.5	Overview of the Process-based Writing attractor
Tab. 6.6	Extracts from the D' Class syllabus (writing)
Tab. A.1	Data generation phases
Tab. A.2	Overview of primary data
Tab. A.3	Coding schemes used for axial coding

List of Figures

Fig. 3.1	The state space of the language school
Fig. 4.1	Affordance landscape
Fig. 4.2	Affordance landscape (Grammar)
Fig. 4.3	Affordance landscape (Vocabulary)
Fig. 4.4	Affordance landscape (Skills)
Fig. 4.5	Distribution of text types across programmes
Fig. 5.1	Dynamics of intentions at the language school
Fig. 8.1	Embeddedness of system
Fig. A.1	Overview of preliminary analytical work

List of Frequently Used Abbreviations

CEFR	Common European Framework of Reference
CLT	Communicative Language Learning
CST	Complex Systems Theory
ECCE	Examination of Communicative Competence in English
EFL	English as a Foreign Language
EIL	English as an International Language
ELF	English as a Lingua Franca
ELT	English Language Teaching
ESL	English as a Second Language
ESOL	English for Speakers of Other Languages
L1	First, native, or mother language
L2	Second or foreign language
KPg	*Κρατικό Πιστοποιητικό Γλωσσομάθειας* [State Certificate of Language Proficiency]
NEST	Native English Speaking Teacher
PCK	Pedagogical Content Knowledge
PPP	Presentation-Practice-Production
RP	Received Pronunciation
SLA	Second Language Acquisition
TESOL	Teaching English to Speakers of Other Languages
TEYL	Teaching English to Young Learners
UMI/ELI	University of Michigan / English Language Institute
WE	World Englishes

Για τη Ναταλία και την Πένη

1 Motivations, background, and queries

The origins of this book go back to a time when I was involved in the management of a language school in Greece. The school had an excellent reputation, which was sustained, among other things, by a strict monolingual policy, and an impressive record of preparing learners for language certification examinations. But what was bizarre, to my eyes at least, was the fact that the school seemed to be entrenched in traditional, teacher-fronted, grammar-focused ways of instruction, which had been evolving in an organic, glacial manner since the 1970s. From the perspective of a person like me, who had been trained in a communicative tradition, such methods could not be effective, and if learners were succeeding, that must have been *despite* the instruction they received, not *because* of it. I was wrong, and this book represents both the process that I went through to realise as much, and an apology.

With an amount of confidence that was only partly due to youth, I took it to myself to 'bring the school forward to the 21st century' (such was the language of ambition in those pre-millennium years). This was a major multi-year endeavour, which involved designing new courses from scratch, providing extensive professional development for the staff, and holding endless meetings where we discussed the whats, hows and whys of communicative language teaching. My own memory of these events was that this had been a futile struggle, although colleagues who have read this manuscript insist that changes were taking place.

I eventually moved on to different things, but I was back in the school several years later to do fieldwork for my doctoral studies. When I returned to my former workplace, one of the first things I noticed was how little long-term impact all the curricular reform had had on the workings of the school. The syllabuses for the courses I had co-created were still in use in many cases, but they had been re-interpreted to fit with the older *modus operandi* of the school. New courses had also been designed, and many aspects of these seemed to align to traditional, transmissive ways of teaching rather than to the principles I had tried to instil. I was also surprised to see that the school staff had developed creative ways of teaching that appeared to conform with the communicative principles that underpinned the curricular review, but on closer inspection aligned to established teaching and learning patterns.

On a personal level, I found this very frustrating, but this frustration was crowded out by a plethora of puzzles that popped in my mind: *Why is the school resistant to reform? What are the processes that forced it to bounce back to its pre-*

ferred state once I stopped applying pressure towards change? If there is no one co-ordinating these processes of 'resistance', why do they seem as if they are somehow organised? It was perhaps a happy coincidence that at the time, I was gradually developing an understanding of complex systems, and I intuitively knew that many of the questions I was asking could be answered by drawing on insights from Complex Systems Theory (CST).

This book, then, is an account of how I attempted to answer these questions, using CST to describe the school. In the book, I draw on data from my PhD thesis (Kostoulas, 2015a), but I am taking a different approach. My objective is not to describe the school as such; rather, I want to provide readers with a template for what a complexity-informed description might look like. In the chapters that follow, I will focus on different components of this description, such as the state space of the system, the affordances that are present in it, the intentionalities that emerge from it, and its attractors. In doing so, I want to show that a description that brings these elements together can provide us with a coherent theoretical account of the phenomena we observe in language education.

As I describe the school, I would like to invite you to perhaps think of similar systems from your own experience (these could be classes that you teach, schools in which you have been, teacher associations, or anything of the like), and to draw parallels between my description and the system or systems that you have in mind. Ultimately, this book will have succeeded in its aim if the conceptual tools that I used to make sense of my experience also help you to make sense of your contexts. And if the overlap is less than perfect, the book will have succeeded in an even more ambitious purpose, that of initiating a discussion on how to best use CST in order to describe phenomena in language education.

1.1. Complex systems in education

1.1.1. What are complex systems?

Before we start the discussion of the school, I think a brief explanation is in order: what exactly are complex systems? This is a topic that is discussed at some length in Chapter 2, but a brief explanation may also be useful here. Complex systems are groups of entities that are so closely intertwined that it makes most sense if we try to understand them as a whole. Entities within a complex system are still ontologically discrete; so, for example, a class or a school is made up of individuals with discrete identities, histories, aspirations, behaviours and so on. But there is also a sense in which the system operates collectively: a system such as a school has a collective identity, a shared history, a common future-oriented trajectory and a collective behaviour. These properties pertain to the system as an entirety, and cannot be reduced to the properties of the people who make it up. In fact, they also have permanence that transcends individual membership. That is, the collective properties of the system stay in place even when one or all the individual constituents have been replaced. Thinking again of the school as a complex system, it may well have a history, traditions and visions, all of which transcend generations. So, when we decide to study a system, this means shifting our focus from the individual constituents to the system as a whole.

What is particularly interesting about complex systems is that they very often behave in ways that are quite unexpected. Characteristically, the activity of complex systems is non-linear. This means that small events could produce disproportionately large effects in the activity of a system, a property popularised in lay usage as *the butterfly effect*. And yet, complex systems can also be surprisingly resilient. So, if an outside influence perturbs their structure or their activity, complex systems tend to reconfigure themselves and maintain equilibrium. It is on account of this resilience that they are also known as complex *adaptive* systems. But while complex systems seem to have preferred patterns of activity, they also change dynamically (hence 'complex *dynamic* systems' or 'complex *dynamical* systems'). This often involves multiple processes of change occurring at different timescales. For instance, if we view an individual's psychology as a complex system, moods can change rapidly, whereas personality traits develop at a slower pace. Finally, and perhaps most importantly, the activity of complex systems is emergent. This means that the interaction between the constituents of the system can produce higher-order phenomena, which are not centrally designed. The emergence of cognition from the bioelectrical activity of the neurons that make up the brain is a typical example.

In writing this brief overview, I have deliberately chosen examples from a range of domains, including neurobiology, psychology and the social organisation of education. I have also drawn examples from a range of levels, from the neurons to entire communities. My intention was to show that CST can help us to understand diverse phenomena that interest us in language education. Therein lies, for me, one of its greatest appeals. Whatever our specific focus in language education, CST can provide us with a unifying discourse, which can help to bring theoretical coherence to the field.

1.1.2 Describing complex systems

Describing an entity like a school as a complex system can seem like a deceptively simple task. One set of difficulties with which we will be immediately confronted relates to how we might define the system – *where do we draw the lines around the system?* When thinking of a school, these lines seem to suggest themselves by the real world: schools have very unambiguous topological perimeters, in the form of building walls or schoolyard fences that separate the school from the community in which it is embedded. They also have recognisable temporal boundaries: we can, for example, decide to concern ourselves only with what teachers and students do between the start and the end of the school day, or maybe our focus could be what happens to a child between their first and last day at school. But even with the clarity provided by such unambiguous limits, it is hard to know what to do with contextual influences, such as decisions by policy makers, or the activity of the students' parents. It is equally unclear how to account for societal attitudes and expectations, which are shaped outside the school, but are embodied by the teachers and students. One might argue then, that the borders of a complex system are suggested by the physical and social worlds, but these borders are somehow fuzzy (Byrne & Callaghan, 2014).

Another set of difficulties connects to group membership. Thinking again of the example of a school, there is an illusion of unambiguousness. Teachers are either members of the faculty or they are not; students are either enrolled in a course or they are not. We might then be forgiven for thinking that we can enumerate the components of a system by consulting the organigram, payroll and student register. The problem with this line of reasoning is that a complex system is not just an aggregate of individual students and teachers. It also includes learning resources, syllabuses, buildings and desks, which may (or may not) need to be included in the description. It even includes non-concrete entities like attitudes and expectations, ideologies and traditions that need to be accounted for. Most importantly, it includes a dense web of relationships that connect all these entities with each other. There is a reason why we call these systems *complex*.

But if we cannot even agree on how to define a system, and what its components are, how can we presume to be able to study it? The answer to this question takes us back to the definition of systems in the previous sections. We focus on describing the system as a whole, not with reference to individual constituents. This is something that will perhaps become clearer in the chapters that follow, where I describe a language school as a complex system. In presenting this description, I will not concern myself with an exact specification of the system's boundaries or an inventory of what components that I believe can be included in its definition. What I will describe, instead, is the outcomes of their interaction, or the phenomena that I observe in the collective activity of the school. In such a holistic description, borders and inventories are of secondary importance.

In the process, I will demonstrate that a holistic description of a school should consist of at least four themes. The first of these themes is what I call the **state**

space of the school. A state space is an imagined outline of all the activity possibilities that are available to a system. This will be further explained in Chapter 3, but for the time being, you might want to think of the state space as a list of all the potential modes of teaching and learning.

The second theme is the **affordance landscape**. Affordances are action possibilities that are available to a system. For instance, if the students and the teacher in a class all have grammar books with them, they are likely to have a grammar lesson. The potential for a grammar lesson is an affordance generated by the book (or, perhaps more precisely, by relation between the class and the learning materials). Sometimes the resources in the system combine in such a way that makes a specific action possibility likelier than others. An affordance landscape is a visual metaphor that depicts the affordances collectively present in the system and the likelihood that they are translated into action. As I explain in Chapter 4, the description of the affordance landscape(s) present in a school can give us insights into teaching and learning practices that take place there.

The next theme in a holistic description of a school is its **dynamics of intentions**. The term refers to a web of different drivers that are present in the system. Drawing on ecological theory, I use the term 'intentionalities' to describe these drivers. In lay terms, intentionalities are the system's purposes. As will be seen in Chapter 5, teaching and learning a language can be associated with many different intentionalities, some of which are only implicit, and all of which are associated with different effects. These intentionalities keep reconfiguring themselves in dynamic ways, and each of these configurations is called a dynamic of intentions. Understanding the dynamics of intentions, I will argue, is necessary for making sense of how the system evolves.

The final theme of the school's description is its **attractors**, i.e., the preferred ways of teaching and learning in the school. In Chapter 6, I argue that these attractors emerge from the interaction between affordances and intentionalities, and they materialise in the state space. In this sense, the description of these preferred pedagogical patterns serves a lynchpin function of connecting all the other elements of the description. In a much more practical sense, it also provides us with a theoretically robust way to account how practices are routinized in the day-to-day operations of a language school.

In the paragraphs above, I listed the four themes with reference to a language school. This was done in an attempt to foreshadow the discussion in the chapters that will follow. However, the thesis of this book is that this tetraptych of state space, affordance landscape, dynamics of intentions and attractors is a necessary and sufficient component of complexity-informed description of many different kinds of systems. This book serves to exemplify how it can be done, in the hope that it might inspire similar work. As you make your way through the pages that follow, you may want to test whether these conceptual tools fit other complex systems that are close to your interests.

1.2 Setting the scene

Before I launch into a complexity-informed description of the language school, in this section I provide some background information about the Greek ELT context. This serves the practical aim of introducing aspects of the local culture(s) that are perhaps unfamiliar, and facilitate the interpretation of passing references throughout the text. At the same time, the overview of context is grounded on the ontological principle that a language school (viewed as a complex system) is integrated with its environment; and that, while it may be pragmatically necessary to distinguish the system from its environment for the purpose of studying it, contextual awareness is still necessary for understanding how the system operates.

A very salient feature of foreign language education in Greece, which is perhaps idiosyncratic, is the co-existence of two language teaching systems that operate in parallel: English is taught extensively in the state-run schools that students attend in the morning, and in the afternoons and evenings the same students often attend additional courses in private foreign language institutes. Although state and private education are different in many regards, and are most easily described separately, this is in some ways an artificial analytical distinction. Students, for example, tend to be members of both systems, carrying their knowledge, expectations and learnt practices from one system to another. Moreover, as will be seen in Section 5.5., whenever there are structural changes in one sector, these are often mirrored in the other. What this means, for the purposes of this study is that, although the language school that I am describing was part of the private education sector, the influence of the state education system needs to be accounted for. So, in the interest of a full contextualisation, state-run and private language education are briefly described in Sections 1.2.1 and 1.2.2, paving the way for a description of the school itself in Section 1.2.3.

1.2.1 ELT in state education

The Greek state education system consists of three main tiers (I am excluding pre-school education and the tertiary sector from the description, as these were not directly relevant to the language school). The first tier, primary education, encompassed three main types of schools at the time when the study was conducted (Table 1.1). Comprehensive Reformed Curriculum schools («Σχολεία Ενιαίου Αναμορφωμένου Εκπαιδευτικού Προγράμματος») were large schools with a staff of generalist and specialist teachers, and the capacity to teach a large range of school subjects. In these schools, English started at Year 1. During the first two years, English classes took place twice weekly, according to the provisions of the pilot Teaching English to Young Learners (TEYL) programme («Πρόγραμμα Εκμάθησης Αγγλικών σε Πρώιμη παιδική ηλικία») (Karavas, 2014). From Year 3 onwards, English courses took place with a frequency of four sessions per week. In addition, many schools provided supplementary English tuition in the afternoon. There were 800 such schools in operation in 2010, and the number was subsequently increased to 961.

'Regular' primary schools, which focused on 'core subjects' such as first language education, mathematics and science, were generally staffed by six to ten generalist teachers and small numbers of specialists whose services were often shared by several schools. In these schools, English classes were held three times per week from Year 3 onwards. Supplementary afternoon classes were also be offered in some schools, depending on the availability of staff. Schools in rural areas («Ολιγοθέσια») were staffed by fewer than six generalist teachers, and they only had the capacity to teach 'core subjects', such as Modern Greek and mathematics, to mixed-age groups. As a result, English was not normally taught in these schools.

Table 1.1: ELT provision in primary education

	Primary (Rural)	2010–2016		2016 onwards
		Primary (Regular)	Primary (Reformed)	Primary (Mainstream)
Year 1 (Α' Δημοτικού)	–	–	2	1
Year 2 (Β' Δημοτικού)	–	–	2	1
Year 3 (Γ' Δημοτικού)	–	3 (+ afternoon?)	4 (+ afternoon)	3 (+ afternoon?)
Year 4 (Δ' Δημοτικού)	–	3 (+ afternoon?)	4 (+ afternoon)	3 (+ afternoon?)
Year 5 (Ε' Δημοτικού)	–	3 (+ afternoon?)	4 (+ afternoon)	3 (+ afternoon?)
Year 6 (Στ' Δημοτικού)	–	3 (+ afternoon?)	4 (+ afternoon)	3 (+ afternoon?)

Since then, the primary education system has been restructured, and the total provision of English has been reduced to one hour in Years 1 and 2, and 3 hours from Year 3 onwards, but this provision has been extended to what used to be 'regular' schools as well. Comparing the curricula of the various school types, one notices a trend towards providing ELT instruction in the state system at increasingly younger ages, which was maintained even in the face of cost-cutting measures. In Section 5.5, we will see how this trend indirectly influenced the language school.

The second tier of education, a compulsory three-year junior secondary school («Γυμνάσιο»), followed a uniform curriculum throughout the country. With regard to ELT, at the time when the study was conducted, learners were divided in 'advanced' and 'false beginner' streams. The allocation of students to streams was done on the basis of their performance on a screening test that usually consisted of reading comprehension, grammar and vocabulary components. The streams were taught separately, but (somewhat paradoxically) both streams were merged in Form 3 and followed the 'false beginner' curriculum, the expectation being that teachers would then apply the principles of differentiated instruction to cater for different levels of linguistic proficiency among the students. English language classes took place two or three times per week. Progression from one form to another was conditional on satisfactory academic performance, which was largely determined by at least two written tests during term time and a formal examination at the end of the academic year. The format of all these examinations was regulated by law: they had to consist of one reading comprehension task and four grammar awareness tasks, and –additionally for Forms 2 and 3– a dictation task.

Table 1.2: ELT provision in secondary education

	Lower secondary	Senior secondary (academic)	Senior secondary (vocational)
Form 1 (Α' Γυμνασίου)	3		
Form 2 (Β' Γυμνασίου)	2		
Form 3 (Γ' Γυμνασίου)	2		
Form 4 (Α' Λυκείου)		3	2
Form 5 (Β' Λυκείου)		2	2
Form 6 (Γ' Λυκείου)		2 (+ project)	2

The third tier of education comprised senior secondary schools, which were either academically or vocationally oriented. When this study was conducted, English was compulsory in both types of schools, although this is no longer the case. ELT classes were held twice or three times per week in academically-oriented schools («Γενικά Λύκεια»). In addition, final-year students were also expected to engage in a collaborative research project, and submit a report in English, although it is unclear how uniformly this legal requirement was implemented. In vocational schools («Επαγγελματικά Λύκεια», «Επαγγελματικές Σχολές»), English for Special Purposes courses were held twice a week. In both types of schools, student progress was monitored by at least one written test during term time and a formal examination at the end of the school year. Similar to lower secondary schools, there was a legal stipulation that these tests comprise tasks assessing reading comprehension, grammar awareness, and writing skills.

The aims of ELT provision in state education were set out in the Interdisciplinary Comprehensive Curricular Framework for Foreign Languages («Διαθεματικό Ενιαίο Πλαίσιο Προγραμμάτων Σπουδών Ξένων Γλωσσών»), and the English Language Analytical Curriculum («Αναλυτικό Πρόγραμμα Σπουδών Αγγλικής Γλώσσας») (Pedagogical Institute, 2003). According to these documents, the aims of English language teaching were to foster literacy, multilingualism and multicultural awareness. In practice, curricular objectives were operationalised in textbooks, which were commissioned by the Ministry of Education, or (exceptionally) selected from an authorised list of commercially available courses. Despite such attempts at regulation, the low specificity in the curricular documents and *ad hoc* commissioning practices resulted in a very low degree of coherence in the learning materials used.

In the state education system, English language courses are taught exclusively by English language specialists, who have graduated from a four-year academic course in English language and literature. This programme of academic studies can sometimes provide a comprehensive grounding in linguistics or North American and British literature, but the provision for language teaching education is minimal (Alexiou & Mattheoudakis, 2013; Kostoulas, 2011; Mattheoudakis, 2007). In-service training for language teachers has been found to be inconsistent in its aims, as well as ineffective in attaining them (Karagianni, 2012). As a result, the most salient influence in the professional development of English language teachers in the state sector appears to be the apprenticeship of observation (Lortie, 2002). Consistent with this hypothesis, it seems that their linguistic and pedagogical beliefs are largely conservative. For instance, it has been reported that they place a high value on the standard language and native-speaker pronunciation norms (Sifakis & Sougari, 2007), as well as teaching grammar (Nikolaidis & Mattheoudakis, 2008). Similarly, research suggests that they tend to favour transmissive and behaviourally informed teaching (Agathopoulou, 2007), while older research, which appears to be still valid to date, has suggested that they placed a premium on grammatical accuracy (Hughes & Lascaratou, 1982).

On the whole, the state-run education system is typified by a largely conservative outlook, with a strong emphasis on the written modality and grammatical accuracy. The influence of this outlook on the language school is discussed in Section 5.5. Although it would be unfair to make sweeping judgments about the quality of the state education system, it is widely perceived as ineffective, in part due to its inconsistent aims, low coherence and unimpressive educational outcomes. This perception has fuelled the growth of a vibrant ELT private sector, which is described next.

1.2.2 The private ELT sector

The private ELT sector is a substantial industry in Greece. According to a recent study, tuition fees for non-state language education was slightly short of 0.9 billion Euros in 2013 (GSEE, 2014), the bulk of which reflected expenditure for English language courses. Available figures indicate that the number of private schools offering evening courses in English exceeds 7,000 (Mattheoudakis & Alexiou, 2009), which suggests that there is one such school for approximately every 400 households. These range in size from institutes with nation-wide presence to small one-person, one-classroom enterprises. Their ranks are swelled by a substantial number of unregistered operations, and teachers offering tuition privately.

Although the private ELT sector is not academically regulated, the language programmes offered by these schools are remarkably consistent. Table 1.3 presents a typical programme of studies, which has been constructed by drawing on my own professional experience of the context, publicity materials, and literature sources such as Mattheoudakis and Alexiou (2013). When I was doing fieldwork for my PhD, students generally enrolled in private language courses at the age of eight or nine, i.e., at about the age when they had their first ELT courses in the state school curriculum, or slightly prior to that. There was, at the time, a growing tendency to enrol at even younger ages, in response to the introduction of ELT lessons in Years 1 and 2 at the 'reformed' primary schools (see above), but this trend appears to have been counteracted by the prolonged economic depression.

Table 1.3: ELT provision in the private sector

Class	Typical age	Language level	Contact hours/week	Notes
Pre-junior	6–8	–	1–2	Optional
Junior A	7–9	A1	3	Often merged
Junior B	7–10	A1	3	
A Senior	9–10	A1+	3	Sometimes offered as intensive summer courses
B Senior	10–11	A2	3	
C Senior	11–12	A2+	3–4	
D Senior	12–13	B1	6	a.k.a. 'Pre-Lower', 'Pre-FCE'
E Senior	13–14	B2	6	a.k.a. 'Lower' or 'FCE'
Proficiency 1	13–15	C1	6	
Proficiency 2	14–17	C2	6	

The majority of students attend six or seven year-long courses ('classes'), which led to certification at the B2 level of the Common European Framework of Reference, or CEFR (Council of Europe, 2001). Since the reputation of schools depends, in part, on the efficiency with which students are prepared for certification, sometimes these courses are conflated. Initially, three contact hours are offered per week, but this provision is usually doubled for older students in exam preparation courses. When I did fieldwork for the study, between a third and half the students who completed the B2-level courses (the 'Lower' class) extended their studies by enrolling in advanced language courses (the 'Proficiency' class). In Section 5.2, it will be seen that this credentialist ethos, which has typified Greek attitudes towards education, is relevant to the discussion of the motivations sustaining activity at the language school.

In addition to being differently structured, private ELT schooling differs from the state education system in terms of the emphasis placed on teaching and learning conditions, especially those that are associated with educational excellence. Classes are usually smaller: whereas state-run schools cap classes at 25–28 students, in private language schools it is highly uncommon to have classes with more than ten learners. Tuition is much more intensive: sessions span 50 minutes to an hour, rather than 35–45 minutes, as is the case in state education. In addition, it is estimated that approximately one fifth of the contact hours outlined in the state school curriculum are cancelled in any given year for various reasons (Pedagogical

Institute, n.d.), but is not the case in private education where replacement lessons are expected by paying parents. Moreover, in the private sector, learning materials are frequently updated, and educational technology is used more extensively than is the case in underfunded state schools. Lastly, private language schools are more sensitive to the societal demand placed on certification and work closely with international certification boards to cater to such needs. The ability of private ELT schools to compare favourably against state education was an important aspect of their culture, not least because of the marketing implications associated with these conditions, a point that we will revisit in Section 5.5 with reference to the language school.

1.2.3 The school

The language school which is described in this book was a large private language teaching institute («κέντρο ξένων γλωσσών» or «φροντιστήριο»), situated in a town in Greece. As is typical for the private ELT sector in Greece, the school mainly provided afternoon and evening language classes, which supplemented the ELT provision in the state education system. The school was distributed in three buildings, located in different areas of the town in order to maximise its catchment area, but these three 'branches' were not in any way organisationally autonomous.

In terms of staffing, there was a cadre of senior teachers, many of whom had a long employment history there, in some cases exceeding twenty years. Most senior teachers were full-time staff, and they were often assigned responsibilities additional to teaching, such as course co-ordination or materials development. Their ranks were augmented by a fluctuating number of teachers on fractional fixed-term contracts. The number of these teachers and their workload varied greatly, depending on student enrolment and their own personal circumstances. Many of the senior staff had been licensed to teach on the strength of entry-level qualifications, which they supplemented with their considerable teaching experience. Younger staff members tended to have additional formal qualifications, in part due to a change in hiring policies intended to modernise the school curriculum (see above).

When I was doing fieldwork, there were between 300 and 500 students enrolled at the school (I have been asked to withhold the precise numbers, as this is considered commercially sensitive information). The majority were young learners and adolescents, between eight and sixteen years old. There were also a large number of young adults, who tended to enrol in exam preparation courses, and a relatively small number of Very Young Learners (under eight years old) enrolled in pilot programmes targeted at that age group. The majority of students had Greek cultural backgrounds, and tended to come from middle-class families.

The school offered a variety of English as a Foreign Language (EFL) courses with differing length and objectives, but the mainstay was the General English programmes, a series of consecutive year-length courses leading absolute beginners to mastery of the English language (Table 1.4). The first of the General English programmes was the *Junior* programme, which spanned two years and catered to the needs of absolute beginners or false starters between the ages of nine and ten. This was followed by a three-year *Senior* programme, which led to the Waystage level (A2) of the CEFR. When the study was conducted, the *Senior* programme was being redesigned: an older version of the programme was being phased out and replaced by what I will call *Senior (New)* (the actual designations referred to the main coursebook used in each programme). One of the main motivations for this change was the trend of students to enrol at increasingly younger ages (See Section 5.5.3 for more details). There was a concern that old Senior programme was proving to be challenging to the youngest of learners, whereas the new Senior programme was considered linguistically, cognitively and thematically more appropriate for that age group. The next programme, Upper Intermediate, consisted of two years of study, and led to certification at the Vantage level (B2) of the CEFR. The final level, which was officially designated 'Advanced Studies', but was more commonly referred to as 'the Proficiency', spanned two years and led to certification at the Mastery level (C2) of the CEFR.

Table 1.4: Overview of the General English programmes

Programme	Class	Age	Target level
Junior	Junior A	9	A1
	Junior B	10	A1
Senior (New)	A Senior	10–11	A1
Senior	B Senior	12	A1+
	C Senior	13	A2+
Upper Intermediate	D Senior	13–14	B1+
	E Senior ('Lower')	14–15	B2+
Proficiency	Proficiency 1	15+	C1
	Proficiency 2	15+	C2

1.3 Outline of this book

The aim of this book is to present a complexity-informed description of the language school, and as I mentioned in Section 1.1.2., there are four necessary and sufficient themes that make up such a description: the state space of the system, its affordance landscape, its dynamics of intentions and its attractors. This list of themes provides me with the organisational frame through which this book is structured.

Before I start the actual description of the school, I introduce CST in Chapter 2. The chapter begins by tracing the origins of complexity thinking, and connecting CST to foundational theoretical and empirical work in mathematics, physics and biology, which has taken place from the mid-19th century onwards. This is followed by a concise outline of CST. In this discussion, I juxtapose CST with more established epistemological paradigms, making reference to three core scientific tenets, which I define as the reductive, predictive and integrative premises. Building on this, I then take a more in-depth look into the properties of complex systems, and derive from them a set of principles that can be used to collectively define CST. In the remainder of the chapter, I discuss how complexity has been used in domains affiliated to language education, such as applied linguistics, education theory and the psychology of language learning, and I suggest some possible ways forward.

In Chapter 3, I begin my description of the school as a complex system by defining the contours of its state space, i.e. the imagined space that encompasses all the potential activity in the school. This is done by means of a visual metaphor, in which the state space is conceptualised as a three-dimensional space. The three axes that I use to construct this space are a linguistic dimension, a pedagogical dimension, and a political dimension. Within each of these dimensions, I define a number of positions, which suggest themselves from the professional literature of ELT. I also note that there are regions in the three-dimensional space that are defined by the cross-section of different positions: for instance, linguistic views that valorise the standard language seem to co-exist in the literature with transmissive pedagogical practices. Based on this observation, I postulate that pedagogical practice might be constrained in these regions.

Chapter 4 moves us to the second theme of a complexity-informed description, the affordance landscape. Affordances, as mentioned in Section 1.1.2, are specific action possibilities that present themselves to the system, and the affordance landscape is a visual metaphor that depicts their strength, or the likelihood that specific action possibilities materialise. I examine affordances in relation to the learning materials used at the school, and use quantitative methods to explore how the content of the coursebooks and assorted materials might influence teaching and learning practices. For reasons of space and analytical convenience, the discussion is limited to the pedagogical affordances (although the methods and constructs that are used in this description could also be extended to the discussion of political and linguistic affordance landscapes as well). Looking at the ways in which affordances are distributed in the system, I note the existence of three types of variation, depending on the activity type, the level of instruction and the year in which each syllabus was created.

Then, in Chapter 5, I look into the dynamics of intention that emerge in the school. Key to this discussion is the construct of intentionality, which is akin to a 'purpose' that drives the system (Kostoulas & Stelma, 2016, 2017). The chapter begins with a definition of intentionality, which I view as a collective, nested, emergent and generative phenomenon (see also Stelma, Onat-Stelma, Lee & Kostoulas, 2015). Next, by synthesising the perspectives of teachers and learners, I trace the emergence of five different intentionalities in the school. These include an imperative to certify linguistic proficiency, an integrative impulse, a commitment to transmitting Anglo-Saxon cultural symbols, a competitive ethos and a protectionistic policy that aimed to safeguard the vested interests of local ELT. Throughout the chapter, the description of intentionalities is interwoven with theoretical insights that are exemplified by the data. The chapter concludes by synthesising these intentionalities into 'dynamics of intention' (Young et al., 2002), i.e. webs of interacting intentionalities. This discussion is connected to the patterns of variation that were teased out in Chapter 4.

In Chapter 6 I describe the actual teaching and learning practices at the school. These practices, I argue, tend to be routinised into predictable patterns. By drawing on interview and questionnaire data as well as lesson plans and syllabus documents, I outline seven prototypical instructional sequences, which seem to act like templates for instruction. Using the language of complexity, I describe these preferred ways of teaching and learning as attractors, and connect them to the state space that was defined in Chapter 3. In the chapter, I also take an in-depth look at three of these sequences (a sequence for teaching reading and vocabulary, a transmissive way of transmitting grammar, and a communicatively-oriented way of developing writing skills), and show how they relate to the affordances and intentionalities that were described in Chapters 4 and 5 respectively.

The last substantive chapter of the book (Chapter 7) brings all the themes together. In this chapter, I take a step back from the description of the school and turn my attention again to complex systems theory and the ways in which it can be used to describe phenomena in language education. Using the description of the school as an illustrative example, I suggest that a complexity perspective can provide a theoretically sound way of dealing with the 'fuzziness' of the systems in which we are interested. I also discuss some of the insights that were generated using the conceptual toolkit that I used in this description to argue that CST can provide researchers with a descriptively powerful and theoretically generative conceptual frame.

The book concludes with a methodological appendix with offers a brief description of the study on which the description of the school is based (for more information, see Kostoulas 2015a: Ch. 4). The appendix begins with a brief methodological positioning of the study. This is followed by an outline of the steps involved in gaining access to the school, generating and analysing the data. This is intended to provide readers with an example of how established research methods, derived from ethnography and grounded theory, can be combined in a complexity-informed investigation.

2 About complexity

My objective, in this chapter, is to trace the outlines of Complex Systems Theory (CST, or simply complexity). In this discussion, I will assume that you, the reader, are not yet familiar with any of the theory or the underlying concepts. It is perhaps unfortunate that complex systems theory had tended to use an array of challenging terms – *sensitive dependence on initial conditions*, *non-linear causality* and more – which can sound intimidating, and which sometimes obscure the fact that they index surprisingly simple, and useful, concepts. What we will do, together, is wade through this terminological quagmire, and at the end of this journey I hope that you will have a working understanding of what complexity is, an appreciation of how this theory might be used to describe aspects of language education, and perhaps even an interest in experimenting with these ideas to make sense of phenomena that interest you.

Some of the points that will be made in the pages that follow have also been raised outside CST, particularly by ecological theory and ecological psychology (e.g., Bronfenbrenner, 1979, 1989; Gibson, 1979; Young et al., 2002). The complexity-informed perspective that I am developing in this book does not make any claims to either supersede such contributions, or challenge the insights that they have provided into English Language Teaching (e.g., Kramsch, 2008; van Lier, 2004a). If anything, it is closely aligned to these theories. Similarly, the discussion that follows is not intended as a critique of more established ways of making sense of the social world, against which it stands as an alternative and complementary view. That having been said, in the pages that follow I will highlight the differences between complexity and other paradigms, and I will point out what their limitations are and how complexity can open exciting new possibilities. My intention in doing so is not to find fault in the alternatives, but rather to trace the contours of CST as visibly as possible.

As we make our way together through Complex Systems Theory, we will try to avoid getting entangled in any of the ongoing debates regarding its ontological premises and epistemological challenges that it poses. If you wish to follow up on these debates, you might want to refer to the excellent treatises by Byrne & Callaghan (2014), Cilliers (2001) and Morin (2006), or to my own doctoral thesis (Kostoulas, 2015a), bearing in mind the old cartographers' admonition: *Here be dragons!* We shall stay clear of such mythical beasts in this chapter, as we make our way through a short description of how CST developed, a tentative definition, an outline of the characteristics of complex systems, an overview of how complexity has been used in the disciplines that concern us, and – finally – a discussion of possible future directions.

2.1 Establishing a new discipline

The theoretical antecedents of CST can be traced to a series of scientific breakthroughs that took place independently from each other in the second half of the previous century. The first of these developments was the publication of the *Outline of General Systems Theory* (von Bertalanffy, 1950). Writing in the context of biology, von Bertalanffy challenged the view that individual organisms can be studied by segmenting them into their constituent parts and then aggregating our observations. In place of this reductive outlook, he proposed focussing on the relationships that connect the constituent parts to the whole organism. Later, in the 1960s, the mathematician René Thom began studying non-linear changes in the behaviour of systems, which he eventually formalised as Catastrophe Theory (Thom, 1972, 1983). A key premise of Catastrophe Theory is that, even though systems are usually resilient, when they happen to be near a 'tipping point', even minor perturbations can have disproportionate effects (e.g., when an already listing boat capsizes).

Another line of thought that contributed to the development of CST was Chaos Theory, which was popularised by publications such as Gleick (1987). While Thom was working on theoretical mathematics, the meteorologist Edward Lorenz was experimenting with computer simulations of weather patterns. Since computers in the 1960s had limited processing power, Lorenz had to work with simplified systems consisting of few elements and straightforward rules, but he discovered that even such systems could behave chaotically, i.e., in ways that were neither truly random nor fully predictable. In a serendipitous chain of events, he also discovered that minuscule changes in the input of his simulations could produce dramatic differences in their output, a property that was subsequently termed as *sensitivity to initial conditions*. His observations were first presented in a paper evocatively titled *Does the flap of a butterfly's wings in Brazil set off a tornado in Texas?* (Lorenz, 1972), from which the popular term 'butterfly effect' has been derived. The impact of Chaos Theory on complexity has been such that some authors occasionally use the terms 'chaos' and 'complexity' interchangeably (e.g., Larsen-Freeman, 1997), but in the interest of clarity it may be best to view chaotic behaviour as just one aspect of how complex systems might behave.

Even more recently, Ilya Prigogine studied what he called 'dissipative' systems, i.e., systems that create internal structures by absorbing energy from outside sources. Crucially, he argued, such systems develop their structure through a process of *self-organisation*, rather than external design (Prigogine & Stengers, 1984). Prigogine's research, for which he received the Nobel Prize in Chemistry in 1977, was instrumental in generating the conceptual toolkit of complexity, and he is credited by Mitchell (2009, p. 298) for creating 'a number of concepts that deal with mechanisms that are encountered repeatedly' in the empirical world, as well as technical vocabulary that has been extensively used in CST, such as 'bifurcations', 'equilibrium', 'homeostasis' and other terms.

The parallel developments that contributed to the birth of CST culminated in the foundation, in 1984, of the Santa Fe institute, an interdisciplinary research centre dedicated to the study of complex phenomena in the natural and social world. In addition, recent years have seen the establishment of interdisciplinary journals focusing on the study of complex phenomena (e.g., *Complexity*, and *Emergence: Complexity & Organization*), as well as the publication of theoretical treatises, such as Byrne (1998), Byrne and Callaghan (2014), Cilliers (1998), and Reed and Harvey (1992), which have outlined the developing theory. Within education, there have been edited collections on complexity including Mason (2008) and Osberg and Biesta (2010). In the domain of Applied Linguistics, *Complex Systems and Applied Linguistics* (Larsen-Freeman & Cameron, 2008) constitutes a landmark publication. More recently, special issues focusing on complexity in *Applied Linguistics* (2006) and the *Revista Brasileira de Linguistica Aplicada* (2013) evidence increasing interest in the potential of complexity to inform research and pedagogical practice in ELT-related fields.

2.2 But what is complexity?

In the absence of a single authoritative definition of complexity (Mercer, 2014), we will develop one such understanding as we go along in this chapter. Pending that, I shall provisionally use the following definitions as starting points. According to Mitchell (2009, p. 4), complexity:

> …seeks to explain how large numbers of relatively simple entities organize themselves, without the benefit of any central controller, into a collective whole that creates patterns, uses information, and in some cases evolves and learns.

Similarly, Nicolis (1995, p. xiii) suggests that the aim of what he calls 'nonlinear science' is to generate concepts and techniques that are required for:

> …a unified description of the particular, and yet quite large, class of phenomena whereby simple deterministic systems give rise to complex behaviours with the appearance of unexpected spatial structures or evolutionary events.

What both these definitions tell us is that complexity concerns itself with the holistic study of complex systems, i.e., systems that are made up of simple constituents but behave in unexpected ways. It is understood that the system constituents are intertwined to such a degree that separating them would be both empirically challenging and analytically unhelpful (Rosen, 1987).

The challenge of defining complexity is further compounded by terminological inconsistency in the literature: terms such as Complexity (Biesta & Osberg, 2010; Bogg & Geyer, 2007; Byrne, 2005; Cilliers, 1998; Rasch & Wolfe, 2000), Complexity Theory (Byrne, 1998; Byrne & Callaghan, 2014; Johnson, 2009), Complex Systems Theory (Larsen-Freeman & Cameron, 2008) and Dynamical Systems Theory (de Bot, Lowie, Thorne & Verspoor, 2013; Valsiner, 1998) have all been used to refer to very similar ways of understanding the natural and social world. It is probably true that such distinctions can highlight theoretical nuance, but my concern is that their potential to generate confusion is greater than any analytical affordances they might provide. In this book, therefore, I shall only use the terms complexity and Complex Systems Theory (CST), and I will do so interchangeably, for reasons of stylistic variety, without implying any theoretical distinction.

The multiple accounts of complexity mentioned above share a theoretical common ground: they all bring into question three foundational assumptions that have underpinned scientific thought from the Enlightenment onwards. I have called these the reductive premise, the predictive premise and the integrative premise.

The **reductive premise** holds that any phenomenon can be best understood if it is disassembled (or 'analysed') into its constituents, which are then studied separately. This principle is explicitly evoked in the *Discours de la Méthode*, where the second and third steps of the scientific method are described as follows:

> De diviser chacune des difficultés que j'examinerais, en autant de parcelles qu'il se pourroit […] de conduire par ordre mes pensées, en commençant par les objets les plus simples et les plus aisés à connoître […] jusques à la connoissance des plus composes. (Descartes, 1637 / 1966, pp. 69–70)
>
> Dividing each of the difficulties that I will examine into as many parts as possible [and] arranging one's thoughts, beginning from the simplest and most easily understood objects […] until one reaches the knowledge of the most complex.

This epistemological outlook is analytically powerful, and it has afforded considerable advances in the understanding of the natural world. However, there is an increasing awareness that understanding natural and social phenomena involves not just knowing their constituents but also the ways in which these constituents interconnect. With regard to the latter, Cilliers (1998) remarks that 'the analytical method destroys what it seeks to understand' (p. 2). From this, it must follow that there is a need for adopting a more holistic perspective of the phenomena under study, and that our epistemological outlook should be 'at the very least interdisciplinary and probably post-disciplinary' (Byrne & Callaghan, 2014, p. 3).

The second foundational assumption of modern science, which I have termed the **predictive premise**, holds that if we can somehow learn about natural or social entities and about the rules that govern their activity, we should (in principle) be able to predict their future states. This optimistic belief has been famously expressed by Laplace, who postulated that if a hypothetical entity ('une intelligence') could, at any given moment, have knowledge of all natural forces and their composition, and if it were vast enough to compute the movements of celestial bodies as well as 'the tiniest indivisible entity' ('du plus léger atome'), such an intelligence would have certain total knowledge of past, present and future (Laplace, 1814, p. 4).

Such confidence in causal determinism was challenged early on. In 1889, Henri Poincaré demonstrated that the behaviour of systems with three or more bodies cannot be mathematically determined using the laws of classical mechanics. He would later claim that:

> Il peut arriver que de petites différences dans les conditions initiales en engendrent de très grandes dans les phénomènes finaux; une petite erreur sur les premières produirait une erreur énorme sur les derniers. La prédiction devient impossible. (Poincaré, 1903, p. 37)
> It might be the case that small differences in the initial conditions produce great ones in the final phenomena; a small error in the former would produce an enormous error in the latter. Prediction becomes impossible.

A similar claim is attributed by Mitchell (2009) to the physicist James Clerk Maxwell, and though it has proved impossible to verify by following up her references, it seems to indicate widespread intuitive awareness of the limitations of the predictive premise, well before they were empirically demonstrated by Lorenz (1972).

Lastly, the **integrative premise** indexes a belief in a 'coherent metadiscourse that performs a general unifying function' (Cilliers, 1998, p. 113). From this perspective, science can be seen as a quest towards the generation of a single explanatory framework. Examples of such attempts include the Theory of Everything in physics, which seeks to describe the interaction of forces and particles from the sub-atomic level upwards, and dialectic materialism, where one might find claims such as:

> Die Dialektik, die sog. objektive, herrscht *in der ganzen* Natur, und die sog. subjektive Dialektik, das dialektische Denken, ist nur Reflex der in der Natur sich *überall geltend* machenden Bewegung in Gegensätzen. (Engels, 1883 / 2011, p. 159, emphasis added)

The so-called objective dialectics, determine nature *in its entirety*, and the so-called subjective dialectics, that is dialectical thought, is only a reflection of the movement through opposites, which *is valid everywhere* in nature.

Some complexity theorists, such as Mitchell (2009), have also adopted the integrative premise, as shown in the following extract: 'we are waiting for the right concepts and mathematics to be formulated to describe the many forms of complexity we see in nature' (p. 302). Whether such optimism is well-founded remains to be seen, but in the meanwhile we may have to do with an alternative viewpoint, which is articulated by Cilliers (1998), drawing on Lyotard (1984). Cilliers notes that 'different groups (institutions, disciplines, communities) tell different stories about what they know and what they do' (p. 114), and that these diverse narratives defy integration. Rather, he goes on to note, 'the postmodern condition is characterised by the co-existence of a multiplicity of heterogeneous discourses' (p. 114), and attempts to bring them together would be 'impoverishing' (p. 118). The challenge that we therefore attempt to address by using complexity might not be one unifying local understandings by replacing their conceptual tools with new ones; rather it must be one of providing a broad frame that can helpfully negotiate their multiplicity.

Complex Systems Theory, then, represents an attempt to develop the theoretical tools that will help us to move beyond the limitations of the reductive, predictive and integrative premises in order produce understandings of phenomena that interest us. Looking into English Language Teaching, which is the particular focus of this book, complexity helps us to develop understandings that are sensitive to the intricate interconnections in the multitude of intra- and interpersonal aspects that are associated with learning; that account for both the rough regularity and delightful unpredictability of the learning process; and that help to meaningfully connect different narratives without negating their differences.

2.3 Characteristics of complex systems

In the previous section, I described Complex Systems Theory as the study of complex systems. Such a definition might appear tautological, so I will now move the discussion forward, by looking into what complex systems are. The starting point in our attempt to understand complex systems is to look into what a (complex or non-complex) system is. Simply put, a system is a collection of entities that interact in specific ways by virtue of their membership in the system (Juarrero, 1999). A collection of books on a bookshelf is not a system in this sense, because the books do not interact with each other or with the shelf. A school system, on the other hand, is made up of individuals who assume specific roles (e.g., student, teacher) and engage in specific behaviours (e.g., lecturing, requesting permission to speak) which are only meaningful and legitimate within the system.

Looking now specifically into complex systems, these are systems that behave in non-linear ways and are capable of adjusting to changes in their environment. A definition of complex systems should encompass two aspects: the structure of the systems, which is *heterogeneous*, *open* and *nested*, and the systems' activity, which evidences *self-organisation and emergence*, *resilient dynamism* and *non-linear* causal mechanisms. In the paragraphs that follow, I will discuss each of these six properties and tease out of them a number of principles that cumulatively make up a working definition of complexity.

2.3.1 Complex systems are made up of heterogeneous components

The most obvious feature of complex systems is that they usually comprise many different components, or 'agents'. I use the term 'agents' very loosely to include any objects, beings, collective entities or even smaller sub-systems that constitute a system. This property, which is called *heterogeneity*, is usefully illustrated by Larsen-Freeman and Cameron (2008), who point out that a traffic system is a complex system consisting of 'citizens, drivers and policy makers', as well as 'roads, vehicles of various sorts, and traffic laws' (p. 28). What is especially interesting in their example is that the constituents of the complex system are not only very different in physical attributes (e.g., cars of different types and sizes) and in roles (e.g., pedestrians and drivers); they are also different in category membership (e.g., roads and laws). Despite their obvious differences, though, these agents may have unexpectedly similar roles, when viewed from a functional perspective. For example, roads and traffic laws are functionally similar in that they constrain traffic, channelling cars in specific trajectories.

Complex systems usually consist of very large numbers of components, but it is important to note that their actual number is not a defining characteristic of complexity. A commonly cited counterexample is a double pendulum, which consists of only two moving parts, but produces complex activity. Rather, what makes systems complex is the large number of interrelations between the agents, a number that increases exponentially if we take the heterogeneity of the system into account.

When working from a complexity perspective in English Language Teaching, being sensitive to the heterogeneity of systems has a number of important implications. First of all, it means that we need to be alert to how agents, such as the learners that make up a language class, are different. This represents a departure from other research perspectives that focus on the similarities among learners, and attempt to understand collective behaviour by reducing it to averages (Miller & Page, 2007). In one of the first studies that applied complexity thinking to language learning, Larsen-Freeman (2006) demonstrated that the students' learning trajectories varied greatly from learner to learner, and from component to component, a finding that highlights the need for theoretical and methodological sensitivity to this variation.

Secondly, awareness of heterogeneity also involves acknowledging that the activity of a complex system, like a school, is shaped by multiple, very different forces. Tudor (2001) offers an example of an account that highlights the diversity of expectations, perceptions and aspirations that come into play in language education, which he calls 'rationalities'. Again, this is a very different way of making sense of language education, compared to studies that focus exclusively on few, and often arbitrarily selected, constituents.

> **Principle 1**
> *Complexity studies systems by synthesising, rather than reducing, their diversity.*

2.3.2 Complex systems have ambiguous boundaries

Complex systems are collections of interacting constituents, which may or may not have a 'real world' physical presence. Whether the constituents that make us a system are 'real world' entities (e.g., students in a classroom) or constructs (beliefs in a cognitive system), the system itself is a conceptual construct. It comes into being when we decide that we want to study a class of phenomena that are associated with a set of interacting entities. Complex systems are created through the process of 'framing', or boundary definition, which is a conceptual move only loosely constrained by the 'real world'. As Cilliers (2001, p. 141) reminds us, the boundaries of the system are both 'a natural thing' and 'a function of our description'. Since system boundaries cannot be specified with precision, it is advantageous to view the systems as 'open' (Larsen-Freeman & Cameron, 2008, pp. 31–33) or 'ambiguously bounded' (Davis & Sumara, 2006, pp. 94–99).

This property, ambiguous-boundedness or openness, is easier to understand by means of an example. As hinted in Chapter 1, a school may be considered a more-or-less clearly defined system, with a visible physical perimeter (the building walls), unambiguous membership roles (described in the student register and staff organigram) and temporal termini associated with its operation (e.g. the beginning and the end of the school day). But even such a system is constantly permeated by 'outside' influences, as we are reminded by ecologically-informed language teaching research that has been attuned to such contextual influences for quite some time (e.g., Holliday, 1994; van Lier, 1988). Such influences might include beliefs that individual teachers and students embody. While administratively self-contained, schools may also be parts of a broader school system, and they normally operate within a frame of externally defined rules and legislation. Schools are also embedded in societal web, and thus subject to outside expectations about their operation and role.

What this suggests, for language education research, is that while we need to conceptually isolate a system from its context for the purposes of inquiry, it is nonetheless necessary to remain attuned to environmental influences. In this sense, the edges of the system might in fact be usefully conceptualised as interfaces through which the systems connect to their environment (Juarrero, 1999), rather than limits or boundaries.

> **Principle 2**
> *Though it is necessary to 'frame' systems in order to study them, complexity aims to explain how they relate to the environment(s) in which they are embedded.*

2.3.3 Complex systems shape each other

A third structural property of complex systems is that they are 'nested', i.e., they exist within broader structures (Byrne, 2002; Haggis, 2008). Davis & Sumara (2006) illustrate this property with reference to classroom mathematics. In their description, mathematics is depicted as a hierarchy of nested structures, similar to a set of Russian dolls: subjective understanding of mathematical concepts is contained in classroom collectivity, which is – in turn – embedded in curriculum structures and, finally, 'mathematical objects', or scientific constructs studied by the discipline of mathematics. Similarly, Larsen-Freeman and Cameron (2008) describe discourse as a hierarchy of systems. They propose viewing discourse events as being embedded within conversations, which form part of broader information exchanges. Nestedness can also be observed in systems that are more visibly grounded on the observable world. For example, a language lesson could be embedded in a school's curriculum, and a school providing English language tuition could be seen as being embedded in the local educational system and global ELT professional culture.

Implicit in the discussion of nestedness is a tendency to view nested systems in hierarchical terms. Juarrero's (1999) description of top-down 'control hierarchies' in which higher-order structures delimit and constrain the operations of lower-order phenomena is typical of such a conceptualisation. But while a hierarchical understanding is intuitively appealing, it is not the only way to think of nestedness. Cilliers (2001), for instance, describes how complex systems may overlap and interpenetrate each other. Similarly, Reed and Harvey (1992) suggest that higher-order and lower-order systems operate in dialectically interacting layers.

Whether we choose to view nestedness as a hiererchy or as a less structured mesh of overlapping systems, what seems to be more important from a theoretical perspective is to distinguish different levels of activity. Doing so has two implications for our description. First, it alerts us to the fact that systems at different levels operate at different timescales (Davis & Sumara, 2006). Lower-order systems tend do be more dynamic than higher-order ones, which evolve at relatively slower rates. For instance, if we view a person's psychology as a multilayered nested system, we might note that moods are constrained by personality traits, and that the former are more volatile than the latter.

The second implication of nestedness is that complexity-informed descriptions need to account for the ways in which higher- and lower-order systems are interrelated. Byrne and Callaghan use the term *reciprocal determination* to describe this interaction. One aspect of reciprocal determination is that lower-order activity tends to be constrained by higher-order structure. This may be the case when established pedagogical practices constrain the teaching and learning patterns observed in individual lessons. The other aspect of this relationship is that the activity patterns in the lower-order systems can shape higher-order structure. In Kostoulas and Stelma (2016), for instance, we describe how small groups of language learners worked together to perform language tasks, and a specific pattern of behaviour gradually emerged from their interaction; this eventually spread, and became established as a classroom norm for a while. This example illustrates how higher-order system structure can emerge from bottom-up activity, in a process called *morphogenesis*. In practical terms, this means that when using complexity in language education research, we need to both remain sensitive to local behaviour, and to connect it to broader structures, in ways that go beyond top-down determinism.

> **Principle 3**
> *Complexity aims to account for reciprocal influences that develop between a system and the structures in which it is embedded.*

2.3.4 Complex systems produce emergent phenomena

One of the most interesting properties of complex systems is that they can develop their internal structure spontaneously, as well as modify it in response to changes in their environment (Cilliers, 1998; Juarrero, 1999). The property of self-organisation has been described as follows:

> [A]s epitomised in swarms of bees and traffic jams, coherent unities can arise without the presence of a leader. The same seems to be true of many sorts of human collectivities in events that are sometimes described as 'grassroots movements' or, more deprecatingly, 'herd' or 'mob' mentalities. (Davis & Sumara, 2006, p. 84)

With regard to the study of second language education, this seems to suggest is that educational systems have the capacity to modify centrally planned structures in order to adjust to local conditions, or even generate structures that are not produced by central planning.

Self-organised activity within a complex system produces *emergent* properties, i.e., 'unexpected spatial structures and evolutionary events' (Nicolis, 1995, p. xiii) that pertain to the system as a whole, and cannot be ascribed to any individual agent. In other words:

> …as systems acquire increasingly higher degrees of organisational complexity they begin to exhibit novel properties that in some sense transcend the properties of their constituent parts, and behave in ways that cannot be predicted on the basis of the laws governing simpler systems. (Kim, 1999, p. 3)

Reading comprehension is a good example of an emergent phenomenon: It is something that cannot be readily explained with reference to the mechanics of holding a document, eye movement, the typography of the text, or the biochemistry of the reader's brain. All of these activities contribute to reading comprehension, but the latter is somehow qualitatively different. To use an worn-out, and yet apt description, *it is more than the sum of its parts*. Since emergent phenomena differ qualitatively from the structures that produce them, the analytical tools used for describing the latter may not be appropriate for studying the former.

Emergence is one of the most important characteristics of complex systems for two reasons. Most obviously, it is the process through which systems self-organise and come into existence. Equally importantly, though, emergent phenomena are recursive: they feed back into the system and shape it, they entrain lower order phenomena, and they affect the higher-order structures in which the system is embedded.

> **Principle 4**
> *Complexity aims to account for how emergent phenomena are produced within a system, and for the ways in which they feed back into the structures that produced them.*

2.3.5 Complex systems tend to be very resilient

Another property of complex systems is their resilience, i.e., their ability to withstand change, and their tendency to exhibit regular, though not entirely predictable, behaviour (Mitchell, 2009). While complex systems have the potential to behave in many different ways, their architecture tends to constrain randomness (Byrne & Callaghan, 2014; Mitchell, 2009). For instance, the norms of a school, the layout of a classroom and the architecture of a building all work together to restrict the degrees of freedom that individuals might otherwise enjoy during a lesson. Due to these constraints, systems tend to remain within 'preferred' states, and revert to them after being perturbed (Larsen-Freeman & Cameron, 2008; Miller & Page, 2007). Examples of such preferred conditions, or *attractors*, include the healthy body temperature of 36.6° C, some people's daily routines, and institutionalised lesson patterns.

While the resilience of systems may appear to preclude the possibility of change, complex systems are not static. In fact, it is one of their hallmarks that their activity is dynamic. Dynamic change can take place within a single, 'cyclical' attractor (Larsen-Freeman & Cameron, 2008), in which case, the attractor is said to consist of multiple states, through which the system moves periodically. An example of such a cyclical attractor might be a recurring lesson pattern consisting of several phases, such as presentation, production and practice. More dramatic changes, or *phase shifts*, take place when a system is dislodged from an attractor and settles in a different one (Larsen-Freeman & Cameron, 2008), as might be the case when a traditionally-oriented curriculum is replaced by a communicative one. According to Delanda, these processes of dynamical change '[break] the link between necessity and determinism […] making the particular end state a system occupies a combination of determinism and choice' (2005, p. 35).

> **Principle 5**
> *Complexity aims to describe both the resilience of structures in a system and their potential for transformation.*

2.3.6 Complex systems connect to their past states in curious ways

While complex systems behave in reasonably regular ways, their precise activity and future states are quite hard to predict. In language teaching, for instance, a teacher might be able to predict how a lesson will develop with a reasonable degree of confidence. They will likely know in advance how the lesson will begin, what activities will take place, and how and when the lesson will end. But as any teacher will attest, even when the same lesson plan is used, no two lessons are entirely identical, and sometimes outcomes are entirely unpredictable. What this means, in practical terms, is that the study of complex systems cannot yield precise predictions of future states: what it can do, instead, is outline the 'limits and boundaries' within which the system's activity is likely to occur (Byrne & Callaghan, 2014, p. 112).

In thinking about how a system's prior activity influences its current state (or how its present activity influences its likely future states) there are two things we need to be cautious about. The first caveat is that complex systems are overly sensitive to their initial conditions, and sometimes substantial effects might be produced by events that are too small to be noticed (Byrne, 1998; Mitchell, 2009). From a research perspective, this means that the usual challenges associated with operationalising and quantifying variables are amplified. Secondly, we should be alert to the fact that the combined effect of multiple actions is not additive. Rather, each agent's activity may reinforce or cancel out the activity of other system constituents, because these are connected by intricate feedback loops (Cilliers, 1998). Put together, these properties of complex systems mean that their activity is *non-linear*. For better or for worse, non-linearity challenges the interpretative power of theoretical accounts that relate causes and results through unsophisticated versions of determinism.

Acknowledging that complex systems do not conform to the expectations of linear causality does not mean that there are no connections between the systems' historicity and their current states, though. Cilliers argues that in complex systems, the 'past is co-responsible for their present behaviour' (1998, p. 4). This is a claim restated by Byrne and Callaghan, according to whom 'structure-agency relationships involve both current and past actors, whose actions have become sedimented into structures' (Byrne & Callaghan, 2014, p. 111). Drawing on Bourdieu (1998), Byrne and Callaghan (2014) propose that the notion of habitus can explain how systems come to be reproduced through the activity of individuals. Activity, in such a view, stems from 'preconscious orientations to action', rather than rational choice or true agency (Byrne & Callaghan, 2014, p. 113). Such activity, they explain, is constrained by pre-existing structures, but leads to morphogenesis, an unintended and unplanned generation of new constraining structures (Byrne & Callaghan, 2014). In other words, complex systems shape themselves through their activity, which is constrained by their past form.

> **Principle 6**
> *Complexity aims to highlight non-linear relationships between the historicity of a system and its present state.*

In the discussion so far, we have looked into complexity, as an epistemological agenda, and complex systems, as an object of inquiry, in an attempt to develop an understanding of what a complexity-informed perspective can offer to language education research. Building on this, we will now turn our attention to how complexity has been used to inform conceptual and empirical work in the field.

2.4 Applications of CST

Recent years have seen a proliferation of studies that have used CST as a theoretical frame or as a metaphor for understanding various phenomena that are relevant to language education. This broadening corpus of scholarship has included work in linguistics, applied linguistics and psycholinguistics, education theory, and language learning psychology, as well as scholarship that attempted to integrate diverse disciplinary perspectives.

2.4.1 CST in the study of language and language acquisition

One broad strand of scholarship that has employed CST perspectives has looked into the potential of complexity to inform our understandings of language and language acquisition. The first tentative encounters of linguistics with complexity theory can be traced back to the late 1990s. In 1997, Larsen-Freeman published a seminal article outlining complexity (or 'chaos' theory, as she labelled it in that early publication) and suggesting its potential to inform the study of language.

Following this seminal publication, several scholars working within this tradition used CST as a lens in order to describe, or re-conceptualise, aspects of the language system. Hanks' (1996) call to view language as a relational system is one early example of this approach. An example that is perhaps better known is Diane Larsen-Freeman's pioneering work on grammaring, where she advocated moving beyond static descriptions of the grammar system and focussing instead on the dynamic processes through which grammatical structures are deployed to create meaning (e.g., Larsen-Freeman, 2001, 2003). Similar proposals have been made, in discourse analysis, for the study of metaphors, which are conceptualised as emergent linguistic phenomena (Cameron & Deignan, 2006; Cameron & Stelma, 2004). Also of interest is an early study by Niyogi and Berwick (1997) who took a diachronic perspective of language, and turned to CST in order to explain how languages evolve over time. Although this early work seemed to take a piecemeal approach, as it tested the relevance of CST to diverse aspects of language, its cumulative effect was to establish the legitimacy of the new research paradigm in linguistics.

In a parallel set of developments, insights from complexity theory were applied to the study of first language acquisition. This work mirrors studies that used CST to describe aspects of developmental maturation (e.g., Thellen & Smith, 1994), and builds on theoretically interesting parallels between complex systems and language acquisition, such as sensitive dependence on initial conditions, dense webs of interconnections and attractors (de Bot, Lowie & Verspoorm, 2007). As early as 1991, van Geert described how cognitive and linguistic development could be understood through the metaphor of an evolving ecological system. Complex Systems Theory was also used to describe the development of L1 phonology (Mohanan, 1992). Positioning herself explicitly in the complexity paradigm, Peltzer-Karpf (1990) described language acquisition as a process of self-organisation, a theory which she continued to develop in subsequent pubications that explored the interconnections between the brain, the language and the environment (e.g., Peltzer-Karpf, 1994, 1996, 2006, 2010, 2012). Similarly, Hohenberger (2002) used the complexity-informed construct of self-organisation to describe L1 development, and Tomasello (2003) also studied child language acquisition from a complexity perspective. A key insight of CST in this work is that it helps to interpret the non-linear trajectories (or 'chaotic itineraries', to use a memorable phrase coined by Hohenberger & Peltzer-Karpf) of L1 development, which occurs in alternating phases of stability, turbulence and fluctuation, followed by stability in a new state (Hohenberger & Peltzer-Karpf, 2009).

Scholars in second language acquisition were quick to integrate these findings in their own work. Larsen-Freeman (2006) used CST to examine the oral and written output of five Chinese learners of English, and showed that their developmental trajectories evidenced dynamic adaptiveness. Larsen-Freeman noted that each learner's learning path was highly idiosyncratic, and as a result CST was better suited to studying this variation compared to research paradigms that averaged out variability. Another CST-informed study in second language acquisition was carried out by Verspoor, Lowie and van Dijk (2008), who used a case study of an advanced language learner, and concluded that second language acquisition is a non-linear process, typified by bursts of progress and regression. Spoelman and Verspoor (2010) also conducted a longitudinal case study of second language acquisition, which showed that noun phrase complexity and sentence complexity seem to develop in alternating bursts. Polat and Kim's (2014) study used CST to examine how the second language skills of an untutored advanced learner developed. Other complexity-informed studies of second language acquisition include Kortmann and Szmrecsanyi (2012), Larsen-Freeman (2002, 2011), Meara (1997), and Mellow (2006). These studies generally attempt to deal with the variability, or 'messiness', of acquisition by replacing rule-based models with connectionist ones.

Other applied linguists have employed a complexity lens to study specific aspects of second language acquistion. For example, Meara (2006) used CST to describe the development of mental lexica, and – in a different study – used the same theoretical frame to account for language attrition (Meara, 2004). Herdina and Jessner (2002) looked into the development of multilingual proficiency from the perspective of CST. Other aspects of language where CST has yielded useful insights include the development of metalinguistic knowledge (Jessner, 2008; Svalberg & Askham, 2016), the development of the writing skill (Baba & Nita, 2014; Verspoor & Smiskova, 2012) and the development of listening skills (Dong,

2016). The studies listed above have demonstrated the explanatory power and versatility of CST, and pave the way for novel conceptualisations of multiple aspects of language.

2.4.2 CST in education and ELT

A second broad strand of scholarship has used CST as a frame for interpreting the processes connected to learning, including second language learning. This line of inquiry is premised on the idea that the structure of the settings where language is taught and learnt, and the activity that develops in these settings, evidence theoretically useful parallels with the phenomena that interest complexity science. This is an argument put forward, for example, by Burns and Knox (2005), who suggest a complexity-informed research agenda focussing on language classrooms. This call has been echoed by Larsen-Freeman (2016), who makes the case for classroom-oriented research, and points out the potential of complexity-informed methods to inform such work.

Perhaps frustratingly in view of the above, empirical work in this area has been relatively scarce (and this is one of the gaps that this study aimed to address). In the general education literature, some scholars who have used complexity include Davis and Sumara (2006), Haggis (2007) and Mason (2008). In the context of language education, Tudor (2001) suggests moving from what he describes as a 'technological' view of education towards an ecological understanding, and uses a complexity-similar perspective to describe teacher identity, learner identity and context (see also Tudor, 2003). In a similar vein, van Lier (2004a) makes a cogent argument for developing an ecological perspective of linguistics, which is sensitive to complex phenomena like emergence, diversity and dynamical activity. He further recommends shifting empirical focus from objects to relations, and developing increased awareness of context. In these studies, complexity is not always explicitly invoked as a theoretical lens, and it is frequently intertwined with an ecological perspective.

Among the scholarship that draws more explicitly on complexity to describe aspects of language education, one study of note has been conducted by King (2013), who synthesised insights from the domains of psychology, sociolinguistics and anthropology to investigate the tendency of Japanese language learners to remain silent during instruction. Another interesting study was conducted by Scholz (2017), who used CST to examine extramural language learning. Looking specifically into how young native speakers of German used English in massively multiplayer online role-playing games (MMORPGs), Scholz argued that the language learning develops as a complex system that adapts to multiple outside factors.

A different approach was taken by Stelma et al. (2015) in a paper discussing intentionality in language settings. Although the authors (including myself, discretely tucked away among the *et alii*) did not explicitly invoke CST in that paper, intentionality is a phenomenon closely associated with complexity (see Chapter 5 for a definition and extended discussion), and the ecological perspective in which our discussion was framed overlaps substantially with that of CST. In that paper, we compared three language learning settings, which we viewed at different levels and timescales: a collaborative writing task in a Norwegian L2 classroom (Stelma, 2003), a Turkish teacher's first year teaching English in primary education (Onat-Stelma, 2005) and the English Language Teaching culture in Korea (W.-J. Lee, 2010). By juxtaposing the findings of the three studies, we hinted at how nested systems evolve at different rates, but are driven by similar forces.

In addition to these descriptive studies, one particularly interesting development connects to what might be a critical complexity perspective. This is a subset of complexity-inspired research that is particularly attuned to questions concerning values, voice and power. One way in which this critical outlook is manifested is by problematising the relative merits of pre-planned versus emergent curricula. It has been suggested, for instance, that an ecological perspective on education can foster 'a spirit of inquiry and reflection' that 'can erode, and to some extent counteract a deficit of rights and conditions in the democratic infrastructure' (van Lier, 2004b, p. 99). Similarly, Biesta (2010) argues that formal curricula are 'complexity-reducing' and disempowering. Another manifestation of a critical complexity perspective relates to the interaction between educational settings and the contexts in which they are embedded. On this topic, Osberg and Biesta problematise 'what kinds of meaning *are allowed to* emerge in the classroom' (2008, p. 313, emphasis added). They further point out that the 'opening up' or 'narrowing down' of the curriculum involve the exercise of political power, and suggest that:

> …the question of the politics of complexity in education is not only about what we might call the promotion of complexity in education, but also has to do with the reduction of complexity in and through education (Biesta & Osberg, 2010, p. 3)

Although English Language Teaching has a rich tradition of critical thinking starting at least since the publication of Phillipson (1992), this line of critical complexity does not seem to have had much impact yet.

2.4.3 CST in the psychology of language learning and teaching

The next strand of complexity-informed scholarship that we will look at involves the adoption of CST in the psychology of language learning and teaching. Some arguments in favour of a complexity perspective in this domain are put forward by Mercer (2013), who brings together insights from complexity theory, ecological thinking, individual difference research and other aspects of the psychology of lagnuage learning. In this rapidly developing strand of inquiry, CST has been used to describe phenomena connected to the self and identity, motivation, cognition and agency, as well as what have been described as whole-person perspectives.

A key theoretical development in this area has been the application of CST to the study of the self and identity. Although scholarship connected to these two constructs and their interrelations with second language learning has a long history (Kostoulas & Mercer, 2016), the first major publication that explicitly invoked CST as an interpretative lens was Mercer (2011b). As Mercer elaborates in a series of publications that followed (e.g., 2011a, 2012, 2016), CST provides us with the conceptual tools to make sense of the dynamism and stability of the language learners' self-concept, and it also affords a relational view in which the self is made up of relationships. Similar insights have been brought to bear on the study of teacher identities. Henry (2016) employs CST to study the development of language teacher identity, which allows him to discuss how the teacher identities of his participants displayed 'multi-stability' and oscillated between two different identity positions.

Another area in language learning psychology where CST has been productively applied is motivation. For example, Ryan & Dörnyei (2013) point out that studies of the L2 Motivational Self System (Dörnyei, 2005) can be helpfully informed by complexity. In their words:

> In the case of many adult language learners the L2 motivational self systems may settle into extended periods of equilibrium; the challenge for researchers is to understand those phases of change when the system moves out of equilibrium and motivation is (re-)energised. [...] A complex dynamic systems approach that places emphasis on mapping unique variation that would have been considered 'noise' by most other paradigms may enable us do this. (Ryan & Dörnyei, 2013, p. 97)

This is a challenge that has been taken up by Sampson (2016), who reports on a year-long, longitudinal qualitative study on the motivation of Japanese learners of English. In this study, Sampson combines action research methods and a complexity lens to describe the students' motivational development as an emergent process, with characteristics including co-adaptation between learners and their environment, self-organisation and non-linear phase-shifts.

Complexity has also been used in the study of other psychological constructs associated with language learning, such as cognitions and agency. Feryok (2018) uses microgenetic analysis to show how cognitions emerge and self-organise, with reference to data from Aslan (2015), Johnson and Golombek (2011) and Feryok and Oranje (2015). The CST perspective that informs her outlook enables her to study cognitions as they emerge, trace their longitudinal development and understand their contextualised nature. CST has also been applied to the study of learner agency by Mercer (2011c). Reporting on a two-year longitudinal study of a language learner, she points out that the learners' agency emerged from the synthesis of multiple factors, and argues that the complexity lens is helpful in capturing the situated nature of the construct as well as the differentiated ways in which it fluctuates across different timescales. Finally, writing in relation to teacher agency, White (2018, p. 202) notes that CST provides researchers with what she describes as a 'promising, as yet unexplored avenue of enquiry'.

The potential of complexity theory to act as a theoretical frame for language teaching and learning psychology is also exemplified by a study that attempted to generate a 'whole-person' description of teachers' psychology. Saleem (2018) conceptualises teacher psychology as an system comprising cognitive, affective, motivational and behavioural aspects, situated in a context; then, using data from two-case studies of higher education teachers in Pakistan, she attempts to describe emergent 'patterns' generated by these systems. She argues that the complexity frame that she employs enables her to understand these aspects of teacher psychology in a holistic, situated way.

2.4.4 CST as connective tissue among disciplines

The three sections above, which described how complexity has been employed in different domains connected to language teaching and learning, illustrate the descriptive power and theoretical versatility of CST. The fact that complexity-informed descriptions have proved so useful in a number of domains suggests that the theory can function as a kind of meta-discourse connecting the diverse disciplines that study language teaching and learning. This is a point that was not lost among complexity theorists, and a number of attempts have been made in the literature to explore how such connections might be made.

In a seminal publication that aimed to describe CST and demonstrate its relevance to applied linguistics, Larsen-Freeman & Cameron (2008) laid the foundation for much of the work that has been described in the preceding pages. This monograph, entitled *Complex Systems and Applied Linguistics*, begins with an overview of complexity theory and complex systems, and following that, Larsen-Freeman and Cameron demonstrate how CST can be used to inform diverse areas of applied linguistics, such as the diachronic evolution of language, first and second language development, discourse, and classroom interaction. The volume culminates in a discussion of the methodological challenges and opportunities associated with the study of CST. The synthetic outlook of the book, which brings together multiple sub-disciplines of applied linguistics under a shared complexity frame, illustrates how CST can be used to draw connections in the field.

Another important step towards this direction was the publication, in 2009, of the 'Five Graces' position paper (Beckner et al., 2009). This joint publication was authored by ten linguists and applied linguists who took part in a conference celebrating the 60th anniversary of the *Language Learning* journal. In it, the authors advance the argument that language, as a whole, can be conceptualised as a complex adaptive system (as is their preferred term). They further suggest that the acquisition, usage and evolution of language are interrelated processes, which they view as facets of the same system. From this position, it follows that CST can function as a unifying frame that brings together scholarship conducted in the fields of first and second language acquisition, descriptive linguistics and historical linguistics.

A similar suggestion has been put forward by de Bot, Lowie, Thorne and Verspoor (2013), who suggest that Dynamic Systems Theory (as is *their* preferred term) can function as a 'comprehensive theory of second language development' (p. 199). In their view, scholarship in second language acquisition has availed itself of multiple 'middle-level' theories, all of which are compatible with each other and with complexity. Among them, they list Vygotskyan and ecological theories that describe the social aspects of SLA, linguistic theories, and theories of second language acquisition (of which the latter need to be re-interpreted by 'challenging many of the commonly established dichotomies that exist in the literature', p. 211). The integrative work that they propose aims at the development of an overarching complexity-informed theory, in which second language acquisition is viewed as an emergent phenomenon that is 'fractally distributed in time and across sociocultural spaces' (p. 216).

In Kostoulas and Stelma (2016), we aim for a somewhat more ambitious goal. We conceptualise language learning as a network of multiple complex systems. We describe three such systems, a linguistic one, an intentional one and a pedagogical one, although it might perhaps be better to view the overarching system as being open-ended to allow for additional systems, as dictated by empirical or theoretical needs. Each constituent system, we suggest, straddles multiple levels, ranging from the individual to the social. For example, at the individual level the pedagogical system consists of preferred or habitual language learning actions. Viewed at the level of small-groups, it consists of routinised practices, and viewed at the level of societies, it consists of entire pedagogical paradigms. Similarly viewed from the bottom upwards, a linguistic system would consist of competences, linguistic repertoires and languages. The frame that we put forward preserves the disciplinary particularities associated with diverse fields, but at the same time CST helps us to describe how the systems interconnect within and across levels.

A somewhat different approach is taken by Hiver and al-Hoorie (2016), who put forward what they call a 'dynamic ensemble', or frame for conducting and evaluating research carried out within the complexity paradigm. The ensemble consists of nine methodological considerations, which the authors contextualise and exemplify with reference to multiple complexity informed studies. Although Hiver and al-Hoorie do not explicitly articulate a meta-theoretical proposal, the methodological frame that they develop serves a similar function of creating common ground among the scholars who engage in complexity-informed work.

In addition to connecting diverse disciplinary domains, complexity can also serve to bridge the perceived gaps between teaching practice and research in the fields of linguistics, psychology and education theory. This argument was articulated by Mercer (2014), who suggested that complexity-informed empirical work can address the critique of pedagogical theory as being at a disconnect from teaching practice. This line of thinking was further developed in a joint paper by the organisers of the Manchester Roundtable on Complexity and TESOL (Kostoulas, Stelma, Mercer, Cameron & Dawson, 2017). In this position paper, we argue that many of the constructs and conceptual tools that are used in complexity are intuitively familiar to teaching professionals, and that – as a result – complexity-informed theoretical narratives are likely to resonate with the lived experiences of teachers. Just as importantly, familiarity with CST can help researchers to develop the theoretical sensitivity that will enable them to more readily engage with teacher-generated narratives of their professional experiences.

The research reported in the previous paragraphs did not, in all cases, explicitly invoke CST as an informing theoretical backdrop, but its ontological tenets are largely similar, and the findings reported are broadly compatible, thus lending credence to the belief that complexity can function as a flexible connecting theory, which strengthens the links across disciplines and epistemological perspectives, while at the same time preserving disciplinary and theoretical particularity. An explicit articulation of how CST might connect to other domains of knowledge, and to other theoretical perspectives lies outside the scope of this study, but I would argue that it forms a key priority for theoretical work in the complexity tradition.

2.5 Ways forward

As hinted earlier of this chapter (Section 2.2), complexity-informed attempts to understand the social world represent a break with traditional ways of thinking about science. This difference in perspective, along with optimism in the potential of the novel way of thought, resonates in the following comment, from Fredrick Turner's introduction to the edited volume by Eve, Horsfal and Lee (1997):

> [By using complexity,] we can dissolve old procrustean oppositions – ordered and random, for instance – and in the process reinstate useful old ideas such as freedom. New concepts, such as emergence become thinkable, and new methods [...] suggest themselves as legitimate modes of study. (Turner, 1997, p. xii)

In the paragraphs that follow, I selectively discuss and appraise a range of methodological tools that seem to lend themselves for complexity-informed empirical inquiry in language education. These include relatively new research methods, such as simulations, time-series analysis and Retrodictive Qualitative Modelling, as well as familiar ones, like ethnography and case studies.

Computer-assisted simulations, or **computational modelling**, are archetypically associated with early complexity research (cf. Lorenz's weather models). This approach involves using a computer programme to generate a simplified model of a complex system, which consists of semi-intelligent agents, rules that govern their behaviour and a set of initial conditions. By running the programme through several iterations, insights can be generated about the long-term behaviour of a system. A theoretical legitimation of this approach is provided by Sawyer (2005), and extended examples can be found in Mitchell (2009, pp. 145–244). Quantitative modelling has not been extensively used in applied linguistics, although some attempts have been made to model language evolution (Baxter, Blythe, Croft & McKane, 2006; Reali & Christiansen, 2009). Despite the paucity of published research, it has been suggested that 'modeling allows one to prove, at least *in principle*, that specific fundamental mechanisms can combine to produce some observed effect' (Becker et al., 2009, p. 12, original emphasis).

My own personal view is that computational modelling is descriptively powerful and efficient, and that it shows great potential in helping us to understand complex social phenomena, especially when driven by empirically collected data, but I am concerned that its usefulness for theory generation is limited in at least two ways. Because they operate at a remove from the empirical world, simulations are prone to blurring the lines between abstraction and fiction. Hedström's caveat, though not addressed specifically to this line of research, seems apt:

> Theoretical assumptions are often seen as mere instruments that can be freely tinkered with until one arrives at simple and elegant models. [...] An explanatory theory must refer to the actual mechanisms at work, not to those that could have been at work in a fictional world invented by the theorist. (2005, p. 3)

Another limitation of computer-based simulations is that they tend to reduce the agents' activity to simple, rational, deterministic, and unchanging rules. This makes them seem unsuitable to investigating issues of agency within a complex system, or questions like the dialectic interaction between the system's structural and behavioural properties (Byrne & Callaghan, 2014).

Unlike computational modelling, **time-series analyses** have generally used empirical data to study complex phenomena. Time-series analysis involves measuring the values of one or more variables at regularly-spaced intervals, and using graphs to plot them, with a view to discovering regularities over time. The potential of time-series to generate insights into developmental issues has been acknowledged in the SLA literature (Mellow, Reeder & Forster, 1996; Ortega & Iberri-Shea, 2005) but few studies explicitly used this research design in SLA, and usually did so in ways that were unrelated to complexity thinking (e.g., Ishida, 2004). More recently, variants of time-series analysis that are sensitive to the time-scales of a system's activity have been recommended as a way to study complex systems (Larsen-Freeman & Cameron, 2008), and attempts have been made to interpret the patterns identified in time series with reference to complex systems theory (Baba & Nitta, 2014; Mellow, 2006).

Although work in this research tradition is free from the kinds of epistemological caveats mentioned in connection to computational modelling, it is at times less than clear how complexity thinking has contributed to the development of findings, other than by providing a technical vocabulary for the description of phenomena that were uncovered by studying the time series. As was the case with computational modelling, time-series research seems best suited to discussing the empirical epiphenomena of complexity, while its potential for uncovering underlying complex rules that generate them has yet to be conclusively demonstrated.

A different approach to studying complex systems, which addresses some of the limitations mentioned above, involves **Retrodictive Qualitative Modelling** (Dörnyei, 2014). The central premises of this approach are that (a) it is possible to associate particular states in a system with specific 'signature dynamics' from which the state emerged; and (b) that despite their limited generalisability, such dynamics can inform our understanding of different situations. To investigate such dynamics, Dörnyei suggests a retrospective approach that consists of three steps. First, the system is studied holistically in order to identify specific states (e.g., a school might be observed to identify 'learner types'). Next, exemplars of each state are identified and subjected to empirical scrutiny (e.g., students representative of each type are identified and interviewed). Finally, salient system components and signature dynamics are identified, and the trajectory that led to the development of the exemplars is retrospectively traced. By synthesising the strengths of qualitative research and a complexity outlook, Retrodictive

Qualitative Modelling appears to offer considerable potential for theory-building, which researchers in ELT are already beginning to tap (e.g., Aslan, 2015; Chan, Dörnyei & Henry, 2014), despite the theoretical and practical challenges of dealing with causality in a complex system.

Ethnographic methods lend themselves well to the study of complex systems. The common ground between the ethnographic research tradition and a complexity outlook is summarised by Agar (2004) as follows:

> For an ethnographer, what's interesting is the discovery of connections. The goal is to build patterns of many interacting things that *include* what was noticed, not to *isolate* what one was supposed to notice and *measure* it. (p. 16, original emphasis)

Agar goes on to make a compelling case about the compatibility of ethnography and complexity. Among other points, he notes that anthropological research has dealt with issues akin to sensitivity to initial conditions, and that ethnographic challenges, such as the researcher embeddedness or the Hawthorne effect, can be helpfully informed by complexity thinking. Some other arguments put forward in the same article, which includes elaborate metaphors of ethnographic work as a search for fractals and fitness landscapes, are perhaps more peripheral to the present discussion, but overall Agar succeeds in two impressive epistemological aims with a single stroke: that of investing ethnography with an alternative ontology and epistemology, and that of providing complexity-inspired researchers with a well-developed methodological toolkit.

The lack of empirical substantiation is an important limitation of Agar (2004), which has been partly addressed elsewhere in the literature. Gomm and Hammersley (2001), for instance, also discuss the overlap between complexity and ethnography, and demonstrate its potential by reinterpreting previous work by Gomm (1986, 2000) from a complexity perspective. In the same vein, Davis and Lazaraton (1995), Holliday (1996) and Ramanathan and Atkinson (1999) are cited by Larsen-Freeman and Cameron (2008) as examples of ethnographic work in ELT that is compatible with complexity thinking. On the whole, there seems to be considerable overlap between the foundational principles of complexity theory and ethnography, and ethnographic methods of data generation appear especially helpful for studying complex systems.

Another research tradition that can fruitfully inform the study of complex systems in ELT is **case study** research. Byrne (2009), for example, notes the similarities between cases and complex systems, both of which attain an ontological 'reality' when they are defined ('framed') by the researcher. Byrne and Callaghan (2014) elaborate on this point as follows:

> [In case study research] we are dealing with things that are both real and constructed, that are fuzzy realities with complex properties, that have a holistic element whilst being constituted from complex configurations, that are intersected with their environment with boundaries being not the things that cut off but rather the domain of intercommunication. (Byrne & Callaghan, 2014, p. 155)

Despite the intuitive overlap between case study research and complexity thinking, references to complexity-informed empirical case studies in the literature seem to be sparse and mostly relate to fields outside education (e.g., Anderson, Crabtree, Steele & McDaniel, 2005; Bennett & Elman, 2006; Lyneis, Cooper & Els, 2001; Thelen & Smith, 1994). Some rare examples of case studies in language education that have been informed by complexity outlooks include Tudor (2001) and Saleem (2018).

This brief survey of possible research methods is of course not comprehensive. Larsen-Freeman and Cameron (2008), for instance, list diverse methods such as formative experiments, action research and brain imaging among the methodological options available to researchers, and it seems plausible that the only real limits are the ones imposed by the researchers' creativity. To return to Turner's comments:

> What the new science has done, in effect, is to place within our grasp a set of very powerful intellectual tools – concepts to think with. We can use them well or badly, but they are free of the limitations of our traditional armory. (Turner, 1997, p. xii)

However, as was argued in this section, the theoretical affordances of each methodological option are perhaps unequal, and research methods such as case study and ethnography seem best suited to the development of rich descriptions that preserve contextual interdependence and situatedness, and from which theoretical insights might be generated.

3 Exploring the school's state space

If the discussion of complex systems and their properties has not put you off CST for good, we can now begin our complexity-informed description of the language school by exploring what, in the language of CST, is called its *state space*. A system's state space is defined as the entire range of possible states in which a system can find itself (Byrne, 1998; Larsen-Freeman & Cameron, 2008; Thelen & Smith, 1994). There are many ways in which the state space of a system may be conceptualised, and the one we choose depends on the structure of the system, on the one hand, and on our own descriptive needs on the other. In this sense, what follows is an example of how the state space of a language education system *might* be contextualised, not a template to be used when describing all such systems. This seems like an appropriate moment to repeat the suggestion I made in Chapter 1 – as you read along, you might want to consider whether this description fits your own context and how you might modify the description to bring it closer to your needs.

Figure 3.1: The state space of the language school

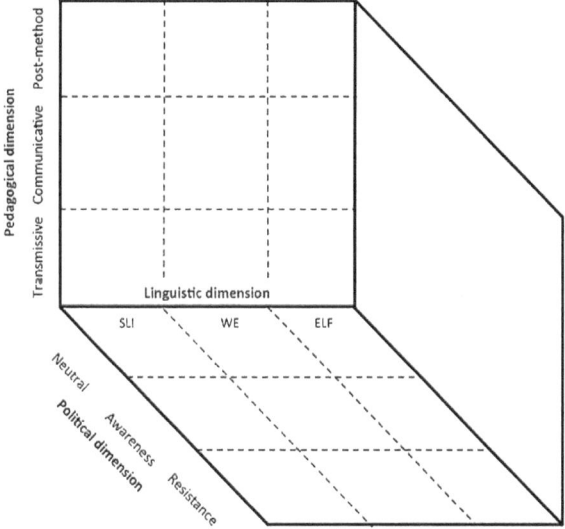

For the purposes of this particular study, I am interested in three aspects of the school's activity: the language taught and learnt, the pedagogical practices, and the political positioning connected to them. Put differently, I am interested in *what* was taught/learnt in the school, *how* teaching and learning took place, and *why* teaching and learning took this particular form rather than any other. This descriptive focus is loosely based on the theoretical framework put forward by Stern (1983), who suggested that making sense of language teaching and learning requires, among other components, a theory of language, a theory of teaching and a theory of society. It is also premised on a personal conviction that the unique role of the English language, seen simultaneously as an instrument and as a product of globalisation, requires broadening the remit of ELT theory beyond the technical aspects of fostering linguistic competence. For ease of conceptualisation, we can visualise the state space of the system as a three-dimensional area, in which each of the three aspects is represented by one axis or dimension (Figure 3.1).

In the three sections that immediately follow, we will look into each of these dimensions or axes. Section 3.1 describes the linguistic dimension by exploring beliefs about English, the language taught in the school. Three positions are identified in this dimension, namely the Standard Language ideology, World Englishes and English as a Lingua Franca. We continue, in Section 3.2, with a discussion of different ways in which language can be taught and learnt. Again, three positions are identified, which correspond to traditional, communicative and post-method pedagogical approaches to teaching and learning. Next, in Section 3.4, I describe three different positions connected to the political dimension of language teaching: neutrality, awareness, and resistance. Having described the three dimensions of the state space, in Section 3.4, I take a closer look into the interconnections between the positions presented in the three previous sections, and trace three regions in the state space of the system, which correspond to established paradigms of language teaching and learning.

3.1 The linguistic dimension: understandings of language

The first of the three dimensions pertains to the concept of the 'target language', and examines three broad options available to the language school. The definition of a 'target language' is a less than straightforward endeavour, especially in view of the global spread of English (Crystal, 2003, 2008; Graddol, 1997; Mufwene, 2001) and the criticisms it has raised (Mesthrie, 2008; Mufwene, 2002; Ngũgĩ, 1986; Phillipson, 2009a; Skutnabb-Kangas, 2000). The multiplicity of linguistic perspectives is represented in this description by critically juxtaposing three linguistic-ideological positions, which are defined with reference to their normative outlook: the Standard Language ideology, World Englishes, and English as a Lingua Franca (ELF).

3.1.1 The Standard Language ideology

The first of the linguistic positions available to the language school is the Standard Language ideology. This is a position that acknowledges linguistic diversity, but holds that it is pedagogically desirable to teach to a clearly defined and socially acceptable standard. Although the 'standard' is often misconstrued as a geographical variety (e.g., British English), it is in fact the *social* language variety associated with the 'social group with the highest degree of power, wealth and prestige' (Trudgill, 2002, p. 124).

An often-quoted advocate of the desirability of teaching the Standard Language is Randolph Quirk, according to whom:

> The relatively narrow range of purposes for which the non-native needs to use English (even in ESL countries) is arguably well catered for by *a single monochrome standard form* that looks as good on paper as it sounds in speech. There are only the most dubious advantages in exposing the learner to a great variety of usage […] all of which is embedded in a controversial sociolinguistic matrix he cannot be expected to understand. (Quirk, 1985, p. 6, emphasis mine)

This argument was re-iterated in Quirk (1990), who criticised recommendations for exposing learners to linguistic diversity in mainstream UK education (Kingman, 1988), and by extension in ELT. While acknowledging the 'idealistic, humanitarian, democratic and highly reputable' motivations of what he termed 'liberation linguistics' (p. 7), he claimed that:

> It is neither liberal nor liberating to permit learners to settle for lower standards than the best, and it is a travesty of liberalism to tolerate low standards which will lock the least fortunate into the least rewarding careers. (p. 9)

Quirk's argument is underpinned by a hierarchical conceptualisation of linguistic diversity, in which *non*-standard forms are assumed to be *sub*-standard, and by the belief that low proficiency in the Standard Language impedes social mobility. This argument reflected widely held societal expectations, which appear to be as true today as they were at the time when the argument was originally articulated (Jenkins, 2007; Williams, 2007).

More nuanced formulations of the Standard Language ideology have been put forward by Peter Trudgill and Alan Davies. According to Trudgill (2002), the Standard Language cannot be reduced to prescriptive rules, and the lines between standard and non-standard forms are blurred by language change. However, he goes on to state, 'from an educational point of view, the position of Standard English as the dialect of English used in writing is unassailable' (p. 127). These remarks were made with reference to first language education, but – as was the case with Quirk (1990) – the reasoning can be, and has been, projected onto foreign language teaching as well. Writing specifically in the context of ELT, Davies (1999) takes a pragmatic stance. In his view, the Standard Language is associated with those whom he describes as 'the educated', and we therefore 'have no choice but to choose or to recommend others to choose Standard English in situations where learners are being educated' (1999, p. 184). However, Davies is perhaps more flexible than Quirk, as he appears sympathetic to the use of those regional varieties that are locally regarded as prestigious.

The Standard Language ideology lends itself to criticism on two accounts. Firstly, it seems to be sustained by narrowly linguistic underpinnings, since it acknowledges the connection between linguistic diversity and social inequality, but stops short of problematising it. Rather, as can be deduced from the paragraphs above, education that is informed by the Standard Language ideology aims at the reproduction, rather than the disruption, of the arguably inequitable social structures in which it is embedded. Secondly, when empirical or theoretical contributions carried out under the Standard Language banner do extend into social problematisation, they tend to do so in ways that are arguably superficial. Pennycook, for one, dismisses the belief that command of a standard variety can bring about social advantages as 'sociological naivety' (2001, p. 48).

Without prejudice to the earnestness of concerns expressed about the liaison between social exclusion and non-standard usage, it seems that the demand for teaching a 'single monochrome standard' is difficult to sustain nowadays on either theoretical or political grounds. In fact, even though defending the Standard Language ideology, Davies (1999) hints at the increasing legitimacy of regional varieties, a point to which I shall turn now.

3.1.2 World Englishes

The global spread of the English language has led to the development of multiple regional varieties, which are increasingly becoming codified and institutionalised (e.g., Adamo, 2007; Kuiper, 2003; Meyler, 2009; Phillipson, 2007). The proliferation of new 'Englishes' has challenged the normative function of the Standard Language. In place of what Widdowson sarcastically describes as 'real English, *Anglais real*, Royal English, Queen's English, or (for those unsympathetic to the monarchy) Oxford English, the vintage language' (2003, p. 35, italics in the original), the notion of World Englishes, 'a pluralised and pluricentric view of English in the world' (Saraceni, 2008, p. 22), in which multiple regional varieties of English have equal standing, has been put forward. This forms the second linguistic position in the description of the state space.

The linguistic legitimation of World Englishes stems from the proliferation of English varieties worldwide. There have been several attempts to theoretically model the diversity and relative status of these new 'Englishes' (McArthur, 1998; Modiano, 1999a, 1999b; Strevens, 1992), of which the most influential seems to be Braj Kachru's (1985) 'three circle' model. Kachru applies historical and geographical criteria to distinguish between three categories of language varieties, or 'circles', as follows: The Inner Circle consists of 'norm-providing' varieties, used in countries where English has historically been a native language. The Outer Circle encompasses 'norm-developing' varieties, which are mainly used in post-colonial polities. These are settings where English performs a significant function (e.g., it is used as a language of education or administration), but is not the native language of the majority of the population. Finally, the Expanding Circle includes 'norm-dependent' varieties, used in the rest of the English-using world. The distinctions that Kachru draws are, at times, problematic (Bruthiaux, 2003; Schreier, 2009), and in light of increasing globalisation it may make sense to redefine the Inner Circle in terms of 'functional nativeness' rather than geographical criteria (Graddol, 2006, p. 110). These limitations notwithstanding, the three-circle model provides us with helpful analytical distinctions, and it highlights the power asymmetries between varieties of English, which World Englishes scholarship seeks to remedy.

In a response to Quirk (1990), Kachru (1991) argued that the Outer Circle varieties should be accorded canonical status alongside the Standard Language. Kachru suggested that concerns about lowering standards ignored linguistic, sociolinguistic and cultural motivations for language innovation. He also claimed that such 'deficit linguistics' (p. 4) overlooked the processes of linguistic institutionalisation that were taking place in post-colonial settings. Further, he challenged the assumptions implicit in Quirk's thesis, such as the belief that English is learnt worldwide for communication with native speakers:

> This, of course, is only partially true. The reality is that in its localized varieties, English has become the main vehicle for interaction among its non-native users, with distinct linguistic and cultural backgrounds – Indians interacting with Nigerians, Japanese, Sri Lankans, Germans with Singaporeans, and so on. The culture-bound localised strategies of, for example, politeness, persuasion, and phatic communication transcreated in English are more effective and culturally significant than are the 'native' strategies for interaction. (p. 10)

Much of Kachru's argumentation is premised on a sharp distinction between the Outer and Expanding circles, and it appears that his main objective was to raise awareness of the legitimacy of Outer Circle (norm-developing) varieties.

By contrast, the norm-dependant status of Expanding Circle varieties was not questioned. However, in view of the blurring of the lines demarcating Kachru's circles that has taken place since the original formulation of the model, the same principles can be applied more broadly than Kachru may have intended (Canagarajah, 2006).

In a similar vein, Canagarajah (1999) challenges the view that non-standard forms constitute evidence of imperfect learning. In his words:

> ...the unilateral movement towards native norms, and the uniform criteria adopted to judge the success of acquisition, ignore the positive contributions of L1 in the construction of unique communicative modes and English grammars for periphery speakers. (p. 128)

Likewise, Yamuna Kachru (1993) draws attention to the dynamism of regional varieties, and calls into question the belief that non-standard forms used by an entire community can be viewed as deficient forms of a standard.

Recent empirical work has confirmed the processes of linguistic change that underpin World Englishes, but the pedagogical impact of these suggestions has not been substantial, especially in the Expanding Circle. In Yamuna Kachru's view, this can be attributed to strong attitudes of antipathy towards regional varieties:

> There are Indians or Singaporeans, who think their own variety is not as 'pure' or 'elegant' as British English [...] This however does not lead to questioning the existence of a standard variety [...] nor should it result in denying the status they deserve to the standard varieties of the Outer Circle. (Y. Kachru, 2005, p. 159)

To provide some examples of such attitudes, Edwards (2010) documents the existence of a regional variety of English in the Netherlands, but finds that attitudes towards it are mostly negative. Similarly, a survey of English language teachers and learners documented that most non-native speakers are more favourably predisposed towards native-like pronunciations and standard grammar than towards non-native equivalents (Timmis, 2002). On the strength of this evidence, Timmis argues that 'while it is clearly inappropriate to foist native speaker norms on students who neither want nor need them, it is scarcely more appropriate to offer students a target which manifestly does not meet their aspiration' (2002, p. 249). Analogous empirical findings have been reported by Murray (2003), Qiong (2004), and Zacharias (2005).

Although the norm-providing function of the Inner Circle varieties appears largely unchallenged (Leung, 2005; Modiano, 2009; Seidlhofer, 2002), empirical and theoretical work in the World Englishes tradition has called into attention the diversity of the English language, and has brought about increased awareness that the selection of a 'standard' for teaching is a politically loaded act. In its drive to advocate equal standing among geographical varieties of English, however, World English has de-emphasised the relations between language and social class, and the way linguistic behaviour is impacted by social context. Another limitation of World Englishes is that as it emphasises codification of geographically defined varieties, it fails to provide an account of the linguistic creativity that emerges in encounters among users of English from different regions. This is the main concern of English as a Lingua Franca, the third ideological position to be examined in this section.

3.1.3 English as a Lingua Franca

In more recent years, the debate about the linguistic content of ELT has shifted to what has been variously termed *English as a Lingua Franca* (Jenkins, 2007; Saraceni, 2008; Seidlhofer, 2011) or *English as an International Language* paradigm (Jenkins, 2006d; Sifakis, 2004; Widdowson, 1997), which constitutes the third position in the system's state space.

English as a Lingua Franca (ELF) is defined as 'any use of English among speakers of different first languages, for whom English is the communicative medium of choice, and often the only option' (Seidlhofer, 2011, p. 7). Within the ELF paradigm, it is claimed, 'all English varieties, native or non-native, are accepted in their own right rather than evaluated against a NSE [Native-Speaker English] benchmark' (Jenkins, Cogo & Dewey, 2011, pp. 283–284). Some scholars (e.g., Firth, 1996; Pakir, 2009; Prodromou, 2008) have used a more restricted definition of ELF, which excludes native speakers, and sometimes 'English as an International Language' is used as an overarching construct that includes communication among both native and non-native speakers. However, since the theoretical potential of these analytical distinctions seems trivial in the context of this study, and considering how inconsistently terminological conventions are used in the literature, we can safely ignore such differentiations in the discussion that follows. In this spirit, I will use the term ELF to index all these connected viewpoints.

The ELF position traces its origins to Jennifer Jenkins' *Phonology of English as an International Language* (2000). In this seminal publication, it is suggested that mutual intelligibility, rather than accuracy (defined as proximity to native speaker usage), should be used as a criterion for pronunciation teaching. On the basis of robust empirical evidence, Jenkins differentiates between 'Lingua Franca Core' phonological features (e.g., most consonants and vowel length distinctions) and features of native speaker pronunciation that do not contribute to intelligibility, such as assimilation, weak forms or the distinction between voiced and voiceless dental fricatives (/θ, ð/). Jenkins argues for de-emphasising the non-core features in pronunciation teaching, and for eliminating the use of problematic native-speaker norms in pronunciation testing. Similar suggestions were reiterated in a number of publications since (Jenkins, 2002, 2006d), and though the pedagogical uptake of these suggestions has been unimpressive, Jenkins' suggestions have enriched the academic and professional discourse with an empirically grounded and theoretically coherent 'demonstration of how things could be done differently' (Saraceni, 2008, p. 21).

Since the publication of Jenkins (2000) there has been a surge of interest in documenting ELF, usually by means of corpus-based research (Kirkpatrick, 2010, 2011; Mauranen, 2003, 2006; Seidlhofer, 2001, 2004). A non-exhaustive listing of ELF features that have been identified so far includes: reduction of 'redundant' features such as third-person singular markers and question tags, shifts in the use of prepositions and articles, and the use of forms with a high semantic generality in place of more specific ones (Jenkins, 2006b; Seidlhofer, 2004, 2011). As regards its pragmatics, ELF communication evidences extensive use of communicative strategies, including code-switching and accommodation, as well as discourse explicitness, repetition and rephrasing (Klimpfinger, 2007; Mauranen, 2012). However, criticisms have been raised regarding the design of the corpora studies and their findings. Saraceni (2008), for instance, questions whether the highly-educated, internationally mobile élites who contributed to the corpora are representative of the majority of English language users. Scepticism has also been expressed as regards the empirical strength of the linguistic claims reported by ELF researchers (Prodromou, 2007), and it has been argued that the distinctions between ELF and communication among native speakers are not as sharp as proponents of the former position make them to be (Sewell, 2013).

In addition to language description and codification, ELF research has aimed at legitimising ELF by researching issues of attitude and identity among English language users (e.g., Dewey, 2012; Kaloscai, 2009; Matsuda, 2003; Sifakis & Sougari, 2005), but the evidence from these studies is largely inconclusive and contradictory. Taking a more critical view, Jenks (2013) points at empirical data that showed that 'the interactants do not see themselves as lingua franca speakers, world English speakers, or speakers of English as an international language' (p. 105), and goes on to argue that the assignment of social categories in research must have at least some relevance to the social interactions and participants under investigation, which does not appear to be the case in much ELF research. While ELF proponents have been quick to dismiss such scepticism as evidence of the 'prevailing "ethos" […] of misinterpretation of and negativity towards the concept of ELF' (Jenkins, 2007, p. 142), their sometimes emotionally charged rebuttals seem uneven in their persuasive power.

A third strand of ELF scholarship has aimed to define and theoretically develop ELF, although it seems that many of these contributions have amplified rather than dispelled the confusion surrounding the construct (O'Regan, 2014; Sowden, 2012a). To begin with, in recent papers it is emphatically stated that ELF is not construed as a specific language variety (Jenkins et al., 2011; Seidlhofer, 2006), but a critical reading of much ELF scholarship suggests that ELF is often reified. For instance, in the context of a discussion of the defining features of indigenised varieties, it is claimed that 'ELF will eventually fit all these criteria' (Jenkins, 2007, pp. 14–15). It is sometimes suggested that ELF is a dynamical language phenomenon (Dewey, 2013; Seidlhofer, 2011), elsewhere ELF is confusingly positioned in the World Englishes paradigm (Jenkins, 2007; Seidlhofer, 2009), and in yet another contribution (Jenkins et al., 2011), ELF is related to 'plurilithic' Englishes (Pennycook, 2007), despite 'a lack of familiarity with the theoretical fundaments of what it means to occupy such a space' (O'Regan, 2014, p. 12). Some ELF scholars (Cogo, 2012; Jenkins, 2007; Seidlhofer, 2006) have attempted to address such inconsistencies, but the credibility of their contributions is by undermined a polemic discourse style, which focuses on addressing perceived grievances rather than on articulating a position, and by a frequent failure to engage with nuances of criticism, which are readily reduced to strawman arguments (Prodromou, 2008; Saraceni, 2008).

When it comes to the pedagogical significance of the ELF programme, recommendations have been scarce (Jenkins et al., 2011) and somewhat nebulous. One line of reasoning is that ELF is a purely descriptive project, and as such does not concern itself with pedagogical prescription. Seidlhofer (2011) is categorical in this regard:

> These chapters [...] are emphatically not intended to prescribe what forms of English people *should* use to ensure effective communication. It is important to emphasise this because the *descriptive* work on ELF [...] has sometimes been confused with the *prescriptive* proposals that have been made for the specifications of a simplified version of English. (p. 154, original emphasis)

Such statements represent a shift away from earlier ELF thinking (e.g., Jenkins, 2000), but this tacit repositioning has generated a certain degree of confusion about the aims of ELF. When venturing into pedagogy, ELF scholars have argued vaguely for moving beyond native speaker norms and for re-educating teachers (Jenkins, 2006a; McKay, 2006; Sifakis, 2004; Sifakis & Sougari, 2005). Strongly deontic arguments are sometimes encountered in ELF discourse, including the call to counter learners' reluctance to embrace ELF as linguistic model, since…

> …ELT seems somewhat bizarrely to be the only educational subject where an important curricular decision (which kind of English should be taught) is seen as being to some extent the prerogative of the students or their parents. (Jenkins, 2007, p. 105)

Elsewhere in the literature, the recommendation is made to raise students' awareness of sociolinguistic variety and provide them a balanced view of alternative linguistic options (Cogo, 2012; Jenkins, 2006c), but even such moderately phrased suggestions seem to extend 'beyond the duty of raising awareness to actual advocacy' (Sowden, 2012b, p. 106).

In summary, the ELF position appears to be a more inclusive form of the World Englishes position, in that it encompasses the Expanding Circle was well. However, closer examination reveals that it seems to be underpinned by a different, though not clearly defined, concept of language: lacking geographical or social anchors, ELF is variously described as an ephemeral product of cross-linguistic encounters, as a reduced variety or as a set of mutually comprehensible varieties. It also departs from the World Englishes position ideologically, in that the liberal *laissez parler* stance underpinning of the former is replaced by what at times can best be described as thinly-veiled linguistic authoritarianism.

Clearly, then, the questions about the target language for ELT cannot be answered in a straightforward manner. Whereas the Standard Language ideology recommends a single, socially defined target variety, the World Englishes position replaces this view with a multitude of regionally relevant targets. The ELF position, it seems, calls into question the very notion of a target language, although as seen in the paragraphs above, it is controversial, both because of the influence of entrenched linguapolitical ideologies and because of its internal inconsistencies. A similar pattern of competing perspectives will be examined in the next section, which focuses on the methods used to teach English.

3.2. The pedagogical dimension: teaching and learning approaches

The second dimension of the system's state space comprises the various pedagogical approaches that were available to the language school. This discussion does not aim to provide a comprehensive listing of ELT methods, for which readers may want to consult Howatt (2004), Richards and Rodgers (2014), and Stern (1983). Rather, the selection and grouping of methods is intended to show the methodological options that were available to the school, either due to their continuing influence on existing practice, or because of their current prominence in the literature. Mirroring the discussion of the linguistic dimension, the pedagogical dimension has been conceptualised as consisting of three positions: the transmissive approach, the communicative approach and the post-method approach.

3.2.1 The transmissive approach

The first position conflates three pedagogical methods that have influenced ELT in Greece, namely the grammar-translation (or traditional) method, the oral method, and the audiolingual/audiovisual method. Despite differences between them, these methods are underpinned by the shared belief that language can be segmented into finite set of words and rules, and that language learning involves sequentially engaging with such components. Although my primary point of reference is ELT, in the discussion that follows I make occasional connections with the informing literature of mainstream Greek education, with which practices at the language school often aligned.

The grammar-translation method is an archetypical form of traditional, transmissive teaching (Howatt, 2004; Stern, 1983). It is underpinned by the belief that language can be reduced to a set of lexical items and grammatical rules to be learnt. Grammatical descriptions provide educators with the scaffolding and building blocks for organising learning events. For instance, lessons may focus on a specific grammar rule or a verb paradigm. Most commonly, grammatical examples are presented in either a decontextualised way, or through literary extracts which are assumed to have intrinsic cultural value. The target structure is described using metalanguage and examples, and there is the expectation that learners should commit this information to memory, and demonstrate knowledge in formal exercises and translation activities. In her derivative outline of ELT methods in the Greek context, Soulioti (2007) uncritically claims that the grammar-translation method was phased out in the 1940s, but this does not appear to be the case, judging by the discernible traces of its influence on ELT learning materials that are marketed in Greece (Prodromou & Mishen, 2008).

The grammar-translation method constitutes an interesting link between ELT and mainstream education in Greece. Grammar awareness has traditionally been an important component of (first) language education in Greek schools, and the emphasis attached to it has been justified as follows:

> Τὸ μάθημα τῆς γραμματικῆς μπορεῖ νὰ ἀποβαίνει μέσον ἐλέγχου καὶ διαρρυθμίσεως τῆς γλώσσας, νὰ καλλιεργεῖ τὴν ἱκανότητα στὰ παιδιὰ νὰ ὁμιλοῦν καὶ νὰ γράφουν ὀρθὴ καὶ σαφῆ γλώσσα […] νὰ ἐφοδιάζει τὸ πνεῦμα μὲ πολλὲς καὶ ὀρθὲς μορφὲς γλωσσικῆς ἐκφράσεως καὶ νὰ διασφαλίζει τὴν ὀρθογραφία τῶν λέξεων. (Kitsios, 1992, pp. 101–102)

> The grammar lesson can become a medium for the control and regulation of language, [it can] cultivate the ability among children to speak and write in correct and precise language […] [it can] equip the intellect with many and correct forms of linguistic expression, and [it can] uphold correct word spelling.

Some features that have typified grammar teaching in mainstream Greek education include (a) the contextualisation of linguistic structures in contrived passages; (b) the pedagogical use of grammar compendia, where traces of classical description survive, albeit in an ostensibly revised form; and (c) the reliance on formal exercises to foster the internalisation of grammar patterns (Vougioukas, 1994). The strongly conservative character of language education in Greece can be attested by the re-introduction, in 1991, of a dated grammar textbook (Tsolakis, 1978) to counter what were perceived as declining language standards, and by the outcry by educators and the members of the public that surrounded its withdrawal in 2011 (Argyropoulos, 2016).

Traces of the grammar-translation method are also very evident in the teaching of classical languages, which form a salient part of the secondary education curriculum in Greece. Teacher training literature (Bezantakos et al. 2008; Revanoglou, n.d.) advocates teaching Ancient Greek using what is described as a text-centred/hermeneutical method («κειμενοκεντρική-ερμηνευτική μέθοδος»). This seems to involve a four- or five-step procedure, which begins with a presentation phase, during which a literary extract is read aloud by the teacher. The reading is followed by intensive reading for comprehension, and a 'synthesis' phase, when students take turns reading the text aloud and translating into Modern Greek, while the teacher explicates problematical lexis and flags grammatical phenomena for further study. The next step, which may take place in a second session, involves focus-on-forms work, during which the teacher describes aspects of morphology and syntax, with reference to substitution tables, tree diagrams and verb paradigms. There appears be a strong tendency among teachers to describe grammar as comprehensively as possible, a practice which is reproved in teacher training manuals going back to Exarchopoulos (1962), and seems to have necessitated tactful reminders that the official curriculum is 'sufficiently detailed' (Bezantakos et al., 2008, p. 13). Finally, students engage with practice exercises that are likely intended to stimulate recall of the formal features of the language. It is to be expected the that the survival of such teaching methods, especially in subjects that are highly valorised in the school system, has a continuing impact on language education, by creating expectations on what constitutes 'proper' language learning.

The oral method is another example of the transmissive approach to ELT. I use the term loosely, to describe a range of similar teaching methods, including the direct method (Howatt, 2004; Stern, 1983), the oral approach (Richards & Rodgers, 2014) and situational language teaching (Howatt, 2004; Richards & Rodgers, 2014). Apart from conceptualisations of language, the common ground shared by these methods is the priority attached to the oral modality. Within this pedagogical tradition, it is considered highly important to expose learners to the target language, or (conversely) to avoid the use of the first language as much as possible. Often, exposure is achieved by presenting learners with texts that have been contrived to contextualise a specific language pattern, and care is taken to select and grade language input so that structurally simple patterns are encountered before more complex ones. Texts are often read aloud by students for practice, with an emphasis on pronunciation, and this reading is followed by display questions, which provide learners with the opportunity to produce language. Lessons often consist of presentation of new structures, followed by controlled practice in the form of drilling, and there is an expectation that, by progressively reducing teacher control, learners will eventually be able to use the target structures in non-classroom contexts.

In this case too, it is possible to draw parallels to mainstream Greek education, as the macro-structure of lessons resembles the 'tripartite' model («τριμερής μέθοδος διδασκαλίας»), a mainstay of primary education in Greece (Chatzidemou, 1988; G. Papageorgiou, 1993). The tripartite model is a simplified version of the teaching model suggested by Johan Friedrich Herbart and his students in the early 19th century (Christias, 1992), which has exerted a formative influence on the techniques and beliefs that make up the teaching culture of Greek education (Matsangouras, 1995). It consists of a progression from information intake to the (re)production of content, which is divided in three stages. During the observation (or acquisition) stage, learners are presented with new content. For instance, a text might be read several times, for holistic understanding and for dealing with challenging parts. Following that, a paraphrase or a summary of the text might be generated by the teacher or students. During the manipulation (or inquiry) stage, students engage with the input in order to create mental representations, and finally during the production (or application) stage, learners are tasked with displaying the knowledge that they produced (Noutsos, 1983). As with the case of grammar-translation, it seems plausible that the prevalence of the 'tripartite' method influences the perceptions of language teachers and learners regarding effective lesson structures.

The final example of the transmissive approach that will be considered here is the audio-lingual method (Richards & Rodgers, 2014; Stern, 1983). As before, the method prioritises the oral modality, and aims at developing fluency through memorisation and mimesis. The method, which is grounded on extensive work in descriptive, structural and contrastive linguistics, as well as behavioural psychology, posits that language can be segmented into small, learnable units, which can then be acquired through habituation and conditioning. Typically, this might involve presenting learners with an example of the target structure (often in the form of a dialogue), which is modelled by the teacher or reproduced from recording. The learners initially listen to the input and repeat the target structures. This is followed by controlled practice (e.g., adaptations of the input text, pattern drills) and, eventually, free production.

The discussion above, which extends in time all the way back to the 19th century *Gymnasien* in Prussia, risks masking much of the diversity and creativity that typified language education. While mindful of this caveat, I believe that the methods described in the previous paragraphs share a number of similarities, which have motivated grouping them in a single position. A key resemblance is that language is, in all cases, understood in formal terms, as a system of rules, patterns or structures that can be taught sequentially. While the theories of learning that underpin each method differ, there are similarities in the ways in which the teachers' and learners' roles are conceptualised: teachers are, in general, responsible for providing or mediating input, whereas the learners' role is, by and large, that of passive recipients of knowledge. The learners' engagement is, for the most part, directed by the teacher, and it usually consists of a linear progression from exposure, through guided practice, to independent use. As we will see later in the book, teaching and learning patterns in the language school tended to align with transmissive pedagogy, which continued to exert an influence on Greek education, even though its foundational assumptions were challenged in the late 20th century.

3.2.2 The communicative approach

Confidence in the scientific principles that informed transmissive pedagogy began to be eroded in the latter half of the 20th century, leading to their complete rejection in the late 1970s. The validity of behavioural psychology, and its relevance to linguistic behaviour was questioned by Noam Chomsky, who convincingly argued that the 'astonishing claims' made by behavioural psychologists were 'far from justified' (1959, p. 49). Not much later, distinctions began to be made between the formal structure of language and its performative functions (Austin, 1962), and between linguistic potential, or competence, and actual performance (Chomsky,

1965). Moreover, empirical work highlighted dialectal variation (Labov, 1966, 1973), hinting that acceptable linguistic behaviour depended on social context. Such thinking was formalised in the distinction between what was formally possible and what was communicatively appropriate (Hymes, 1972). At about the same time, descriptive linguistics shifted focus from formal features of the language to the notions that it conveyed (Wilkins, 1976) and to its functions (Halliday, 1973). In 1980, an influential model of communicative competence was put forward, which distinguished between grammatical, sociolinguistic, discourse and strategic competences (Canale & Swain, 1980). The cumulative effect of all these breakthroughs, alongside widespread frustration with existing methods, was a paradigm shift in language teaching, and the development of the communicative approach to language teaching, which forms the second position in the pedagogical dimension of the system's state space.

In the discussion that follows, I use a loose definition of the communicative approach to encompass a broad range of teaching methods, which primarily use linguistic interaction as a means to develop communicative competence. The key pedagogical principles that underpin this approach to language education are that the ability to communicate is not dependent on pre-existing meta-linguistic knowledge; that communicative competence can be fostered through engagement with communicative tasks; and that learning activities should be meaningful to the learners (Richards & Rodgers, 2014).

The theoretical justification of the communicative approach extends in three directions. First, the belief that familiarity with linguistic form facilitates the ability to use language was challenged. According to Widdowson:

> Knowing what is involved in putting sentences together correctly is only one part of what we mean by knowing a language, and it has very little value on its own: it has to be supplemented by a knowledge of what sentences count as in their normal use as a means of communicating. And I do not think that [transmissive methods used at the time] (make) adequate provision for teaching this kind of knowledge. (1972, p. 17)

Taking a different, but complementary, approach, it has been suggested that exposure to linguistic input is a necessary but not sufficient condition for developing linguistic competence. Rather, it is necessary for learners to have ample opportunities to grammatically encode meaning through the production of output (Swain, 1985). Similarly, the claim has been put forward that actual communication can foster learning, as it provides scope for the negotiation of meaning and conversational adjustment (Long, 1985). A third strand of argumentation involves the affective impact of communicative language teaching, which is argued to be more motivating for learners, and seems to create conditions conducive to learning (Littlewood, 1981).

Some formulations of the communicative approach can be described as task-supported pedagogy (Ellis, 2003). An example of such pedagogy is provided by Littlewood (1981), who advocates the use of 'pre-communicative' as well as 'communicative' activities. The former category includes activities focussing on structure, which familiarise learners with the target forms, and 'quasi-communicative' activities that 'take account of communicative as well as structural facts about the language' (p. 86). The latter category includes functional communication (e.g., sharing, exchanging or processing information) and social interaction (e.g., role-plays, open-ended discussion etc.). These teaching methods are often associated with an attempt to identify, sequence and teach the components that make up communicative competence, using organisational principles that are not too far removed from structural syllabuses (Ellis, 2003).

A different conceptualisation of the communicative approach involves a process-based syllabus, such as the one used in the Communicational Teaching Project (Prabhu, 1987). The project, which took place in schools in and near Bangalore, India, used cognitive tasks as a means for teaching English. The students who were engaged in tasks were required 'to arrive at an outcome from given information through some process of thought', whereas teachers were expected 'to control and regulate that process' (Prabhu, 1987, p. 17). For instance, learners might be tasked with finding and describing locations on a map, or extracting information from a railroad timetable, using English as a medium of communication (Prabhu, 1987). Typically, learning sequences consisted of a pre-task activity, in which the task was modelled by the teacher, and the actual task, which was carried out by the learners. The Bangalore Project, as it is alternatively known, has not been replicated, partly because it was finely attuned to the particularities of its social context, but it has provided impetus for pedagogical developments since.

A more recent outshoot of the communicative approach is the development of Task-Based Learning (Ellis, 2003; Nunan, 2004; Willis & Willis, 2007; Willis, 1996). Task-based learning events are structured around tasks, defined as 'classroom work which involves learners in comprehending, manipulating, producing, or interacting in the target language [...] in which the intention is to convey meaning rather than to manipulate form' (Nunan, 2004, p. 4). Tasks might involve any or all skills, and have a clearly definable outcome (Ellis, 2003). Examples include listing, ordering and sorting information, comparing, problem-solving, sharing personal experiences, or engaging in creative work (Willis, 1996). An influential model, put forward by Willis and Wills (1996) consists of pre-task activities, intended to familiarise learners with the task; a task cycle, including interaction, preparation of a report and presentation of outcomes; and language work, focussing on language features that become salient during the task.

The communicative approach, broadly construed to include diverse variants such as the ones listed above, has attained the status of methodological orthodoxy in ELT, at least in terms of its representation in the informing literature, curriculum documents and official policy. Its key features, such as the primacy attached to meaning as opposed to form, the attempt to simulate authentic interaction within the classroom, and the specification of learning activities with reference to the process rather than their expected outcomes, are expounded in the professional discourse, and they are generally espoused, often uncritically, by educational authorities, publishers and ELT practitioners. However, it is not entirely clear whether actual practice always aligns with these espoused beliefs (Nunan, 2004). In fact, in the chapters that follow, we will witness some traces of the tensions between 'encroaching' communicative influences and persistently surviving transmissive practices. But for the time being, I will confine myself to the observation that recent years have seen an erosion in our confidence that any one method can universally inform practice. Having made this remark, I now turn our attention to a line of scholarship that articulates this scepticism.

3.2.3 The post-method approach

Despite the apparent dominance of the communicative approach, there is ample reason to believe that it has not always been attuned to local educational needs. To mention just a few indicative instances of mismatch, it has been reported that communicative language teaching is incompatible with the Chinese culture of learning (Hu, 2002), and that it places unreasonable demands on overworked and under-qualified teachers in non-western learning contexts (L. Yu, 2001). Elsewhere in the literature, it is suggested that communicative language teaching may be inhibited by a variety of institutional and affective factors (Li, 1998). In the Greek context, it has been noted that language learning materials are, for the most part, incompatible with communicative principles (Kostoulas, 2007; Prodromou & Mishen, 2008). From a more theoretical perspective, there have been calls in the literature for greater sensitivity to contextual factors (Bax, 1997; Richards & Rodgers 2014), and for viewing aspects of the local culture 'not as constraints to be overcome but conditions to be satisfied' (Widdowson, 2004, p. 369). These critical views, which constitute the third position in the pedagogical dimension, are presented below.

The compatibility between communicative language teaching and local context has been a key theme in a series of publications by Adrian Holliday (1992, 1994, 2005). Holliday (1994) puts forward an understanding of learning environments as complex webs of overlapping cultures (e.g., student cultures, professional academic cultures, classroom cultures, and national cultures, etc.), and he argues that it is possible for cultural intrusion to occur when aspects of a 'donor' national culture (such as the Anglo-Saxon West) are incompatible with aspects of the 'host' national cultures in the periphery of the English-speaking world (1994). The evocative metaphor of 'tissue rejection' is used to describe how transplanted teaching practices might fail to take hold because their effect on local settings is disruptive (Holliday, 1992). To mitigate such problems, Holliday advocates using 'appropriate methodology', i.e., teaching methods that are grounded on ethnographically derived understandings of local practice (1994). He argues that communicative language teaching, in the prescriptive sense outlined in the previous section, may be culturally inappropriate in many settings, but the underlying principles of communicative ideology ('treat language as communication', 'capitalise on students' existing communicative competence', and 'communicate with local exigencies') are culturally transferrable and can therefore function as guidelines for the development of appropriate methodology (2005, pp. 143–146).

The concept of 'exploratory practice' (Allwright, 2003, 2005; Allwright & Hanks, 2009) is another way of moving beyond methodological prescriptivism, which (like Holliday's suggestions) foregrounds the importance of sensitivity to local conditions. It is premised on the belief that teaching practice should be informed by universally valid principles, but that teaching practitioners should be mindful of the implications of applying these principles locally (Allwright, 2003). To that end, exploratory practice is put forward as 'an epistemologically and ethically motivated framework for conducting practitioner research in the field of language education' (Allwright, 2005, p. 361). Exploratory practice involves adjusting pedagogical practices to empirically derived understandings of the local context, through a process that consists of identifying 'puzzles', reflecting on them and monitoring them, taking action, considering the outcomes, and publicising one's understandings. Although exploratory practice does not explicitly position itself in the post-method tradition, it seems to share its emphasis on making sense of context and adjusting pedagogy to it.

The post-method macro-strategic framework, developed by Kumaravadivelu (1994, 2001, 2003, 2006c) constitutes a comprehensive set of guidelines and principles that complement the previous recommendations. Central to this framework are the parameters of particularity, practicality and possibility (Kumaravadivelu,

2006c). Particularity refers to the need for pedagogy to be attuned to local exigencies. Practicality is understood as the priority attached to practice-derived understandings that teachers develop. Finally, possibility refers to the aim of affirming personal identity and challenging social injustice. Kumaravadivelu suggests that practice should be informed by ten macro-strategies (e.g., activating intuitive heuristics, promoting autonomy, raising cultural consciousness), which can be operationalised in a number of micro-strategies or teaching procedures (Kumaravadivelu, 2006c). By providing teachers with the flexibility to design and extend context-appropriate micro-strategies, while at the same time offering a set of universal principles that prevent unproductive relativism, it is suggested that this framework offers scope for the development of 'a systematic, coherent and relevant theory of practice' (Kumaravadivelu, 2006b, p. 69).

Despite differences in nuance and emphasis, these three different instantiations of post-method thinking share a number of fundamental assumptions. First, they are underpinned by the belief that the universal application of specific methods is pedagogically inefficient and ethically questionable, if done without reference to local particularities. Moreover, they all involve a distinction between contextually-sensitive ways of dealing with day-to-day teaching, and underlying principles which are argued to be universally relevant. Even though it has been suggested that the procedures described for operationalising these principles can be construed as a type of method (Liu, 1995) – perhaps even a method tainted by Western ideology (Bax, 2003; Canagarajah, 2002) – the post-method approach represents a paradigmatic shift from the approaches described in previous sections, by virtue of its readiness to engage with the social milieu in which teaching takes place. The discussion of social context is carried forward to the following section, which deals with the political implications of ELT.

3.3 The political dimension: visions of society

The third dimension of the school's state space pertains to the relation between ELT and the social settings in which it takes place. This 'political' dimension encompasses multiple issues, such as globalisation and local identities, linguistic and cultural imperialism, the tensions between universal values and cultural relativism, and aspirations of empowerment and equality, all of which had a bearing on the forces that motivated teaching and learning at the language school, even though their relevance was sometimes only implicit. This discussion anticipates themes of native-speakerism (Holliday, 2005), cultural incursions from the Anglophone west and resistance against 'foreign' teaching methodology, which are examined in greater depth in the chapters that follow. Reflection regarding the interface between ELT and social processes can be undertaken from a multitude of perspectives, which I group under three headings: a neutral position, a position of critical awareness, and a position of resistance.

3.3.1 Neutrality: seeing no evil

The neutral position is typified by a narrow linguistic or pedagogical focus on how ELT is understood. As such, teaching and learning that aligns to the neutral position does not concern itself with political and social implications of language teaching. This position is neatly summarised in the following response to concerns raised by Phillipson (2009a) about the spread of the English language:

> What worries me is that, through your metaphors, human qualities are assigned to English which is, after all, *just a language*. It has no *ability* to do things by itself, nor does it bear the *responsibility* for its state of being. (Dendrinos, 2009, p. 181, emphasis in the original)

We shall not concern ourselves here with the rather bizarre definition of language as an entity disconnected from its users, or with the surprising assertion that language cannot 'do things'. Instead, we can use this quotation as an example of an ideological position about language that emphasises preserving narrow disciplinary boundaries. For an interdisciplinary profession like ELT, however, such a position is unhelpful in uncovering the social effects of teaching practices. I will illustrate this point below with reference to the way ELT often deals with the constructs of 'authenticity' and 'culture'.

Narrowly-focused pedagogic approaches to ELT often prioritise 'authenticity', which 'loosely implies as close an approximation as possible to the world outside the classroom' (McDonough & Shaw, 1993, p. 43), either through exposure to genuine (i.e., not pedagogically contrived) texts, or by simulating real-life communication. In the professional literature, one can find multiple examples of arguments for authenticity, as well as descriptions of pedagogical practice that illustrate the use of authentic materials (e.g., Guariento & Morley, 2001; W. Y.-c. Lee, 1995; Peacock, 1997; Wong et al., 1995). An extreme example is provided by Seargeant (2005), who describes attempts to replicate British culture in Japan, by creating learning environments that mimic British architecture and display realia imported from the UK. Staffed with native speakers, such 'English villages' provide opportunities for natural interaction in the target language. Such practices are not explicitly informed by a theory about the language-culture interface, but they seem to conflate the teaching of linguistic form with the unidirectional transmission of cultural imagery and cultural values from the Inner Circle outwards (Nault, 2006; Shin, Eslami & Chen, 2011).

There are also recommendations in the literature for enriching the language curriculum with cultural information, 'culture' being variously construed as awareness of the literary and artistic heritage of the Inner Circle countries, or as information about everyday life in such settings (Byram & Feng, 2004). Sometimes such recommendations are justified on instrumental grounds. For example, it has been suggested that incorporating literature in the language curriculum can increase learner motivation (Ghosn, 2002), and similar claims have been made about other aspects of the target language culture (Dörnyei & Csizér, 1998). Similarly, the explicit study of cultural differences that are encoded in equivalent lexemes across languages has been pedagogically recommended (e.g., Cortazzi & Shen, 2001; Olk, 2002). Whether or not such cultural enrichment practices are pedagogically effective is debatable (Alptekin, 1993; Byram, Esarte-Sarries & Taylor, 1991), but what is definitely problematic is that they often tend to be associated with superficial cultural representations (Aliakbari, 2004; Kostoulas, 2011c) as well as essentialised understandings of the target language cultures (Alptekin, 2002).

The examples above are indicative of either naiveté about the political implications of ELT, or cynical indifference to them, but a 'neutral' positioning might also be informed by ideological aloofness. This stance is best exemplified by Widdowson (2003), who argues that it is possible to be committed to a belief in social justice, while at the same time preserving a boundary between personal convictions and academic work. Widdowson argues for an outlook that looks to linguistics (as opposed to politics) for its 'disciplinary point of reference' (p. 13), and he cautions against the dangers of confusing applied linguistics with 'politics applied' (p. 14). After professing a belief in political agnosticism, Widdowson states that 'the applied linguistics that informs the kind of enquiry I undertake [...] does not impose a way of thinking, but points things out that might be worth thinking about' (p. 14). Implicit in this statement is a distinction between applied linguistics proper, and problematisation about ELT, which is argued to belong to the domain of personal reflection and political activity.

The 'neutral' position that was outlined above stands in stark contrast to understandings of ELT that we will examine next, namely views that explicit engage with the political implications of English language education.

3.3.2 Awareness: developing a critical understanding

A second way of conceptualising the relation between ELT and the social contexts in which it is embedded involves drawing attention to association of language education with processes of linguistic or cultural hegemony.

Central to this debate is Phillipson's *Linguistic Imperialism* (1992), a landmark publication that 'unapologetically takes us out of the sheltered world of language teaching into more uncomfortable realities surrounding that language' (Holborow, 1993, p. 358). The main thesis of the book is that 'the dominance of English [as an international language] is asserted and maintained by the establishment and continuous reconstitution of structural and cultural inequalities between English and other languages' (Phillipson, 1992, p. 47). According to Phillipson, the inequitable distribution of power between what he terms the 'core' and the 'periphery' of the English-speaking world is sustained by ideologies of Anglocentricity and professionalism (the latter term being defined as a narrow focus on the linguistic and pedagogical issues that face ELT). Phillipson further argues that the aggressive promotion of English by UK- and US-based international organisations prioritises resources of the core over those of the periphery, a claim that he backs up with a modest amount of empirical, experiential and historiographic evidence, and considerable rhetorical force.

Unsurprisingly, Phillipson (1992) has generated much controversy. Concerns have been raised, *inter alia*, about the empirical rigour of the book (Berns et al., 1998; Canagarajah, 1999; Davies, 1996; Fishman, 1993), about an unhelpfully antagonistic authorial style (Berns et al., 1998), and about overstatements that risk invalidating possibly legitimate claims (Crystal, 2000; Davies, 1996). While the use of analytical tools such as Gramsci's theory of hegemony enable Phillipson to probe deeply into social reality in ways that had not been previously attempted in applied linguistics, all too often it seems that much analytical prominence is given to instances where language spread can be explained with reference to power asymmetry, at the expense of phenomena that do not fit Phillipson's theoretical model (e.g., the role of English in Scandinavia or Japan, or the spread of Spanish in the USA). Despite such shortcomings, Phillipson (1992) has made a powerful impact on ELT. The linguistic imperialism thesis has been reiterated several times (Phillipson, 2001, 2003, 2004, 2009a, 2009b), the role of English as a global language has been debated (Crystal, 2000; Phillipson, 1999), and empirical data on the relation of English to local linguistic ecologies is increasingly being published, both supporting and refining Phillipson's thesis (e.g., Bisong, 1995; Canagarajah, 2005; David & Govindasamy, 2005; Griffin, 2001; Lin & Martin, 2005; Mesthrie, 2008; Rajagopalan, 2005; Ramanathan, 2005; Randall & Samimi, 2010).

Broadening the debate regarding the social impact of ELT, critically-informed scholarship has called attention to the connections between language teaching and the forcible imposition of values that are alien to local communities. Edge (2003), for instance, decries the ancillary role of ELT in agendas of military conquest and religious conversion. Pennycook and Coutand-Marin (2003), and Pennycook (2005) raise similar concerns about the cultural politics involved in conflating ELT and evangelical work. Strong disapproval of the prosyletisation practices often associated with ELT in some settings has also been expressed by Varghese and Johnston (2007), who question the compatibility of the Great Commission (the imperative, in Christian doctrine, to spread Christianity) with core tenets of education, such as fostering 'doubt and the capacity for changing one's mind' (p. 27). On the other hand, there are voices in the literature for what is designated as 'faith-informed' (or more explicitly: Christian) ELT (e.g., Dörnyei, 2009, Purgason, 2004, 2014; Smith, 2008; K.-M. Yu, 2007), and recent years have seen the publication of edited collections exploring the overlaps between language education, faith and religious identity (Wong & Canagarajah, 2009; Wong et al., 2013). While a discussion of the relative merits of each position lies outside the scope of this book, all these examples indicate an increasing awareness of the conduits for the transmission of cultural values through language teaching.

Kumaravadivelu (2006a) provides a theoretically elaborate account of such conduits. He suggests that the 'dangerous liaison' between globalisation, empire and ELT can be analysed into four intertwined processes. To begin with, ELT promotes western knowledge over local understandings, thus furthering the vested interests of western scholars. Secondly, local languages are displaced from the ELT curriculum. Moreover, Western culture and English are fused with a view to creating 'cultural empathy' towards the inner circle communities. Finally, from an economic perspective, ELT generates income and employment opportunities for members of the Inner Circle communities at the expense of local economies. To counter these undesirable effects, he recommends a philosophical, pedagogical and attitudinal re-structuring of ELT.

Much of the scholarship quoted above seems to construe the globalising flows which sustain ELT (and are sustained by it) as unidirectional. A somewhat different view is presented by Pennycook (1994), who argues that the tensions between local languages and English are greater than what might be inferred by superficial analysis. Pennycook describes the co-existence of two conflicting educational agendas in colonial policy that advocated education in English (Anglicism) and local languages (Orientalism), and notes that:

> First, both Anglicism and Orientalism operated alongside each other; second, Orientalism was as much a part of colonialism as was Anglicism; third, English was withheld as much as it was promoted; fourth, colonized people demanded access to English; and finally, the power of English was not so much in its widespread imposition but in its operating as the eye of the colonial panopticon. (p. 103)

The formative discourses that emerged from these antagonistic orientations to education, he argues, are still traceable in the competing values and attitudes about English ('critical ambivalences') that are encountered in post-colonial settings (p. 74). He also points out that while English has traditionally been associated with globalisation and imperial power, it has also been the language of protest, and can therefore have empowering effects. In making these points, Pennycook shifts the debate from a discussion of social reproduction to a discussion of resistance, thus paving the way for the third position, which is presented below.

3.3.3 Resistance: empowering teachers and learners

The theme of resistance, which forms the third political position in the state space, is picked up by Canagarajah (1999), who uses ethnographic methods to provide an example of resistance-oriented critical pedagogy. Canagarajah notes that narratives such as the Linguistic Imperialism thesis are limited in two ways. First, they are overly deterministic and do not offer scope for the operation of individual agency:

> Such evidence leads us to the conclusion that students come with a relatively independent consciousness that can display signs of opposition to domination; that the cultures they bring to them can clash with alien ideologies to resist domination; that human experience is of sufficient complexity and indeterminacy to override what impersonal institutions may predicate; and that students may enjoy some agency to challenge reproductive forces. (p. 25)

In addition, they seem to overlook the fact that the dominant cultures display diversity, and may contain counter-cultural discourses, on which the disenfranchised may tap to resist reproduction. Canagarajah uses extensive ethnographic examples to illustrate how the imposition of dominant culture was resisted by students in Sri Lanka. In doing so, he adds a valuable empirically grounded perspective to the discourse about the interface between culture, language teaching, and identity.

A similar example of resistance is described by Holliday (2005), drawing on an unpublished paper by Jacob (1996). Working at an Indian university, Jacob and her project team are said to have counteracted the encroachment of cultural values from the Anglophone West 'through a conscious *resistance discourse*' (p. 168, emphasis in the original). This process consisted of a number of strategies, including 'marking territories', 'reconciling with the past', 'speaking against the grain', and 'moving to centre stage from the periphery'. The actions of the project team led to the generation of a counterculture, i.e., 'a small culture which asserts itself against a dominant small culture or cultures, but one which […], tries to make sense of the dominant culture in different ways in order to survive and make itself known' (p. 171). Holliday goes on to argue that ELT practitioners ought to work with such locally emerging countercultures, with a view to bridging them with the dominant cultures.

The ideal of bridging cultures is prominent in Holliday (2005), who proposes 'cultural continuity' as an alternative to difference-oriented understandings of culture. He defines cultural continuity as 'an appreciation of how cultural realities and practices connect and mingle to allow collaborative inclusivity', and goes on to suggest that it can form 'the basis on which all ESOL educators can be equally represented and valued' (p. 157). According to Holliday, developing cultural continuity involves rethinking some of the fundamentals of the language teaching profession, such as the dichotomy between native- and non-native speakers, or the ways in which the periphery is represented in the curriculum. Within this framework, it is suggested that ELT can be usefully deployed…

> …as an instrumental, though very rich resource, which does not have to represent the cultural values of the English-speaking West, but which can absorb and express the depths of whatever cultural values with which it becomes associated. (p. 165)

Though empirically grounded on a number of interviews with English language teachers from over the world, Holliday's proposal seems to be exploratory, rather than a fully worked out example of how cultural continuity might be achieved. It nevertheless constitutes an articulate theoretical legitimation for practically-oriented suggestions, such as the one that will be presented next.

The Multicultural Awareness Through English (MATE) approach to ELT (Fay, Lytra & Ntavaliagkou, 2010) is, in some respects, similar to the proposals outlined above. Described as a paradigmatic possibility for ELT in Greece, MATE involves English-medium communication among members of culturally diverse student communities, with a view to increasing their cohesion. Key to this approach is the development of intercultural space, to which participants bring unique sets of linguistic and cultural resources, and within which they can freely develop their identity repertoire. Because MATE was developed with reference to the Greek educational context, Fay et al. make frequent references to 'pupils' and 'intrasocietal' encounters. However, the underlying principle of equal participation in a neutral social space, akin to the 'third space' of language teaching (Kramsch, 1993, 1995), seems to have relevance to other instances of English-mediated communication. Thus, MATE provides an example of how cultural diversity can be pedagogically exploited in language education for the purposes of empowering learners and promoting equality.

A common ground shared by resistance-oriented scholarship includes rethinking of the culture construct, which is understood as a property of small groups, rather than in reductive, essentialised terms (Holliday, 1999). Seen from this perspective, empowerment-oriented ELT is associated with two agendas: First, it aims to raise awareness of how such groups emerge in the process of language teaching, and how they relate to dominant cultures. In addition to that, it involves using language education to create small cultures where students are empowered.

3.4 Constraining structures

The description of the system's state space is intended to show the range of possible states in which a system can conceivably find itself. However, as we saw in Section 2.3.3, the degrees of freedom that any system enjoys within its state space tends to be constrained by higher-order structures. Thinking specifically of a language school, some examples of such structures might be the prevailing educational norms in the culture(s) where the school is embedded, or the expectations articulated in the professional and academic literatures. These constraining structures constitute one of the elements that predispose a system to behave in specific ways (the others are connected to available resources, as explained in Chapter 4, and intentional drivers, which are discussed in Chapter 5).

Thinking specifically of the language school that is described in this study, its activity seems to be constrained by three higher-order structures: (a) the norms and practices associated with local educational cultures, (b) those associated with global ELT culture(s), and (c) those represented in the critical ELT literature. I will use the terms technical, mainstream, and critical structures, respectively, to refer to each of these. As shown in Table 3.1, each of these structures is associated with a different combination of theories of language, learning and society.

The first, or *technical*, constraining structure stems from the beliefs and practices usually encountered in Greek education. It is associated with the Standard Language ideology, transmissive teaching methods and a lack of problematisation over the political implications of ELT. The *mainstream* constraining structure, thus named because of its canonical status in the professional literature, stems from teacher education and the professional discourse of ELT. Like the technical constraining structure, this is also neutral as regards the political implications of ELT. Another similarity is that is broadly informed by the Standard Language ideology, but it may accord canonical status to certain indigenised varieties of English. However, it differs from the technical paradigm by virtue of its adherence to communicative language teaching, as opposed to transmissive pedagogy. Lastly, the *critical* constraining structure, which stems from the critical ELT scholarship, challenges established theories of language, by replacing the notion of a 'target language' with the notion of competence in a broad repertoire of linguistic varieties. Pedagogical views are informed by post-method thinking, and a by readiness to defer to local expertise. Most distinctively, the critical constraining structure is typified by an agenda that involves raising awareness of the political processes associated with ELT, and empowering teachers and learners.

Table 3.1: Constraining structures in the state space

	Technical	Mainstream	Critical
Higher-order structure	Local education norms	Teacher education Professional literature	Critical ELT literature
Linguistic dimension (Theory of language)	Standard Language	Standard Language World Englishes	World Englishes ELF
Pedagogical dimension (Theory of learning)	Transmissive	Communicative	Post-method
Political dimension (Theory of society)	Neutral	Neutral	Awareness Resistance

This chapter began the description of the language school by describing its state space, i.e., the range of shapes that teaching and learning in school might conceivably take. The state space was conceptualised as a three-dimensional space, defined by a linguistic, pedagogical and political dimension. Following that, I traced the contours of three constraining structures in the state space, which corresponded to specific ways of teaching and learning. This sets the scene for the next step in the description, where I look into the resources available to the school, and examine how these predisposed the system to move towards (two of) these constraining spaces.

4 Tracing the affordance landscape

In this chapter, I take a further step in the complexity-informed description of the language school, by continuing to describe its structure. The previous chapter discussed ways in which higher-order structures, such as established pedagogical norms, professional discourses or literature recommendations, could constrain the shape of teaching and learning that emerged in the school. But these top-down constraints are only part of the system's structure, the other part being the resources and influences that are available to the school. These resources and influences are the focus of the present chapter.

Before moving on, some remarks are in order about terminology. Although the term 'resources' could encompass a wide range of artefacts, from classrooms and equipment to pens, posters and notebooks, in this chapter I will focus only on the learning materials that were used at the school. This will include published materials (i.e., coursebooks and accompanying resources), and materials such as worksheets or tests that were developed in-house. Describing these resources is useful for two reasons: Firstly, they offer some insight into what was considered linguistic and pedagogical appropriateness in the school at the time of study. When this description is combined with the higher-order constraints that were outlined in Chapter 3, together they can account for the existence of attractors within the school's state space. Secondly, if we take into account when and why the materials were introduced, this can also provide some clues about how these perceptions of appropriateness evolved over time. In doing so, this description hints at the diachronic development and dynamism of the attractors.

4.1 Learning materials creating affordances

In the context of this discussion, it should be remembered that the precise relation between materials and practice remains contentious. For example, Jacobs and Ball (1996) have suggested that coursebooks have minimal influence on practice. They suggest that 'some teachers ignore teachers' manuals and even the instructions in the student's book' (p. 101), and opt for lesson planning practices that are more autonomous and responsive to students' needs. In their perspective, the determinants of learning events are the teachers' views on the learning process, as well as the learners' needs, rather than coursebook content as such. This position is consistent both with methodological orthodoxy, which has emphasised the need for teacher creativity, and with empirical findings according to which teachers reportedly prefer to modify coursebooks for specific learning situations (e.g., Leung & Andrews, 2012). While it is not my intention to challenge the role of individual autonomy, or suggest that teachers are insensitive to the needs of their students, I believe that Jacobs and Ball may be downplaying the role of published materials as constraints on agency, and they seem to disregard the ways in which coursebook content shapes expectations among teachers, learners and stakeholders regarding what may be regarded as legitimate or effective lessons (see also Littlejohn, 1998; Nunan, 1991).

A more nuanced position has been taken by Hutchinson and Torres (1994). In an often-cited article, they point out that the published courseware helps teachers to manage learning, and it addresses the learners' need of structure. Hutcheson and Torres reject the strongly deterministic view that learning materials constitute scripts for pre-planned learning events (as suggested, e.g., by Littlejohn, 1992), and propose that they function as frameworks for flexible lesson planning, and as drivers for curricular change. Although Hutchinson and Torres do not articulate a fully developed theory regarding the role of learning materials, their argument seems to suggest that these materials limit the freedom of choice that would otherwise be available to teachers and students, and help them to channel agency in predictable ways.

Without prejudice to the overall validity of the claim put forward by Hutchinson and Torres (1994), empirical data from the Greek context appears to suggest that teachers in Greece tend not to deviate much from the recommendations in the teacher manuals or the courses of action implicit in the design of the activities. For example, Alexandropoulou (2002) provides evidence that teachers in the state sector in Greece tend to plan their lessons with reference to their coursebooks, rather than learner needs or the national curriculum. Similarly, M. Papageorgiou (2002) describes teachers in the private EFL sector as 'course implementers', whose duties are limited to the delivery of the content in the set coursebooks (p. 50). Analogous findings have been reported in other studies, including Georgiadi (2003) and Xanthakou (2005).

What seems to be needed, then, is a theoretical way of accounting for the role of learning materials, which can accommodate the multiple relations between learning resources and pedagogical practice described above. One way to do this would be by drawing on the construct of *affordances*, which is derived from ecological psychology (Gibson, 1977, 1979). I define affordances as action possibilities implicit in the design of an activity, and I suggest that they influence pedagogical practice in two ways. Directly, they privilege certain ways of teaching, by making them easier to implement compared to alternatives. Indirectly, they shape the teachers', learners' and other stakeholders' expectations regarding the 'legitimate' content and format of language lessons. Affordances could relate to any of the three dimensions of the state space, by privileging specific linguistic, pedagogical and political positions. However, the purpose of this chapter is not to present an exhaustive list of all the affordances present in the system; what I want to do instead is to illustrate their role, the ways in which they are created and the influence they exert on practice. To that end, the discussion that follows will focus on the affordances connected to the pedagogical dimension, and will show that how they tended to gear the system towards transmissive and communicative (though not post-method) ways of teaching.

Figure 4.1: Affordance landscape

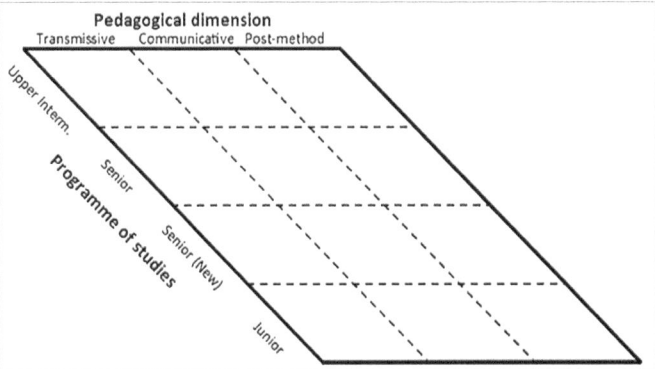

Collectively, affordances construct what can be described as the *affordance landscape* of the system (Figure 4.1). Like the state space, this is a metaphorical visualisation of the collective effect of the affordances that are implicit in the learning materials, and also of the probability that certain forms of pedagogy materialise in practice. The affordance landscape that will be used as the basis of this description is a simplified version of the state space, in which one axis corresponds to the pedagogical dimension of teaching, and the other one represents different programmes of study in the language school. Figure 4.1 shows an entirely level affordance landscape, in which all possibilities are equally likely. However, as will be seen in the next sections, depressions in this plane represent action possibilities that are more likely to happen, on account of the structure and content of the learning materials.

4.2 Overview of the learning materials

In this description, I use the term learning materials to describe a core set of materials that were used in actual teaching, as defined in the school's syllabus documents. Materials, in this restricted sense, comprised published courseware (typically a coursebook and a workbook) and in-house supplementary materials. As I report in Kostoulas (2015a), these were accessed through a combination of systematic and theoretical sampling (Strauss & Corbin, 1994), and analysed through content analysis (Krippendorf, 2004).

The structure of the courseware in the language school was modular. The basic building block was a *course* i.e., a set of textbooks and activity books that were used in a single programme. Each course consisted of two or three *levels*, designed for use in a single academic year. Every level was divided into four to six *modules*, which were further subdivided into two to four *units*. A typical unit, which often cohered around an overarching topic, was made up of three to five *lessons*. Lessons normally took up a double-page spread, and comprised several *activities*. While the structure of individual lessons sometimes varied, the structure of units and modules within each course was identical. A detailed list of the courses, units and activities that were studied and the instruments used for cataloguing their content can all be found in Kostoulas (2015a). In the paragraphs that follow, I will refer to the courses with fictitious names.

Table 4.1: Structure of a typical module (Junior)

Lesson	Focus
Unit 1	
Lesson 1a	Reading (Vocabulary), Grammar
Lesson 1b	Reading (Vocabulary), Grammar
Lesson 1c	Vocabulary, Writing
Unit 2	(as above)
Revision	Vocabulary, Grammar

The course used in the Junior programme (hence *Junior*) was published by a major international publisher expressly for the Greek ELT market. The course was divided into two levels (A and B), which had been adopted for the A and B Junior classes respectively. The course components that were used were the Class Book and the Activity Book. Each level was divided into six modules, consisting of two units and a revision lesson, which tested grammar and vocabulary. Each unit contained three lessons. Of them, two began with a story in cartoon form, which – according to the authors – 'introduces the target [grammar] structures in a clear and meaningful context'. This was followed by a formal presentation of a target language structure, and several listening, speaking and writing activities, where the structure was practiced. The third lesson in each lesson focussed on vocabulary and also offered scope for developing the productive skills.

While I was conducting my study at the language school, a restructuring of the curriculum was in progress, which involved the gradual introduction of the *New Senior* course. The course, which was also published by a major international publisher 'for Senior classes in Greece', consisted of three levels, of which the lowest, *New Senior 1*, was already in use in the first of the three years of the programme (A Senior). Much like *Junior*, *New Senior 1* was divided into six two-unit modules, and each unit contained three lessons, with additional 'revision' or 'fun' lessons interjected between them. The first and second lessons in every unit contained passages (a continuing story in dialogue form and a quasi-genuine text) which contextualised grammar and lexis, and these were followed by grammar presentation and practice activities. The third lesson in each unit focused on writing. Although *New Senior* was generally well received by the staff, not least because it was accompanied by numerous Interactive Whiteboard resources, concerns had been voiced that the materials did not provide enough grammar coverage, and to that end, an additional grammar book was used to supplement the course.

Table 4.2: Structure of a typical module (New Senior)

Lesson	Focus	Lesson	Focus
Unit 1		Unit 2	
Lesson 1a	Reading (Vocabulary), Grammar	Lesson 2a	Reading (Vocabulary), Grammar
Lesson 1b	Reading (Vocabulary), Grammar	Lesson 2b	Reading (Vocabulary), Grammar
Lesson 1c	Writing	Lesson 2c	Writing
'Fun' spread	Speaking	Revision	Grammar, Vocabulary

The *Senior* course, which was being phased out, consisted of three levels, of which *Senior 2* and *Senior 3* were still in use. Much like the courses described above, *Senior* had been produced for the Greek ELT market by a local imprint of an international publisher. Each level of the course comprised four modules, corresponding to the four two-month terms into which the academic year was divided. There were three units and a review in every module, and each unit consisted of four lessons. The first two lessons in every unit were based on dialogues that formed a continuing storyline, and the third one was structured around a quasi-authentic article or similar text. The passages in these 'numbered' lessons served to practice reading skills and present vocabulary; they also contextualised grammar, which was presented and practiced later in the same lesson. The fourth lesson in the unit ('Plus'), which focused on writing, contained a model text and guided writing activities. The *Senior* course was very popular among the teaching staff, who praised its rigorous and methodical structure and its perceived comprehensiveness. However, there was a growing recognition that topical texts were dated, and that the content of the course seemed at times too challenging for learners who began English at an increasingly younger age.

Table 4.3: Structure of a typical module (Senior)

Lesson	Focus
Unit 1	
Lesson 1	Reading (Vocabulary), Grammar
Lesson 2	Reading (Vocabulary), Grammar
Lesson 3	Reading (Vocabulary), Grammar
Plus	Writing
Unit 2	(as above)
Unit 3	(as above)
Unit 4	(as above)
Revision	Grammar, Vocabulary

At the Upper Intermediate programme, the course in use was *Exam*, also published by a local imprint of an international publisher. It comprised three levels, of which *Exam (B1+)* and *Exam (B2)* were used in D and E Classes respectively. Each level was divided into six two-unit modules, and each unit contained five double-page spreads focussing on reading, grammar, vocabulary, the oral skills and writing. For reasons of consistency, I have used the term 'lesson' to describe the double-page spreads, although this was not actually used by the publishers of this particular course. The layout of the book encouraged flexible use: for instance, the reading and vocabulary sections were often taught together, despite the fact that they were physically separated by a grammar section. Although *Exam* was not marketed as an exam-preparation course, its outlook was strongly exam-oriented (hence the fictitious name used here). In the Teacher's Book that accompanied the course, it was explicitly stated that the course 'offers extensive practice of the types of exercises in the Cambridge FCE [First Certificate in English] and Michigan ECCE [Examination of Communicative Competence in English] exams'. Moreover, almost all the activities in the course were modelled after common examination tasks, and 'steps to success' boxes with examination strategies were found in most pages. In addition to the *Exam* materials, collections of practice tests were also extensively used towards the end of the Upper Intermediate programme, both as teaching materials and for developing examination skills.

Table 4.4: Structure of a typical module (Exam)

Lesson	Focus
Unit 1	
'Lesson' 1	Reading
'Lesson' 2	Grammar, Vocabulary
'Lesson' 3	Listening, Speaking
'Lesson' 4	Grammar (exam practice)
'Lesson' 5	Writing
Unit 2	(as above)
Revision	Grammar, Vocabulary

In addition to the published materials, teachers and learners in the language school also used a sizeable corpus of resources that had been created in-house. Many of these materials had been commissioned by the school to supplement perceived deficiencies in the published materials. In addition, a small number had been designed in in-house professional development workshops, and some had been created by individual teachers for their classes, but were shared in the spirit of collegiality. This corpus of in-house materials was made up of four main categories of resources. First, many materials, including practice tests, newspaper articles, podcasts, video-recordings, etc., were available for the Proficiency programme, for which no coursebook had been set. Secondly, teachers and learners were encouraged to use monolingual wordlists, which were used to facilitate vocabulary learning in the Senior and Upper Intermediate programmes. An additional category was made up of photocopiable worksheets, most of which contained grammar notes and exercises, as well as writing activities. Lastly, there were a number of self-contained thematic and occasional lessons, which would be used at the teachers' discretion whenever a deviation from the courseware was deemed appropriate.

As discussed in Section 4.1., the learning materials were associated with specific affordances, and we will now move on to describe those. This discussion is organised in four sections, which correspond to the affordances generated by activities that had a Grammar, Vocabulary, Oral Skills and Written Skills focus.

4.3 Affordances in the grammar activities

Grammar activities tended to create very strong affordances connected to transmissive ways of teaching and learning (Figure 4.2). This effect was mainly generated by the high prevalence of activities explicitly focussing on developing metalinguistic competence and grammatical accuracy. This seemed to be the case particularly in the Junior and Senior programmes, where presentations and controlled practice activities were very common. In the Upper Intermediate programme, however, there was a shift towards more communicative pedagogy, as the coursebook, *Exam*, contained fewer grammar activities, many of which involved inductive learning.

Figure 4.2: Affordance landscape (Grammar)

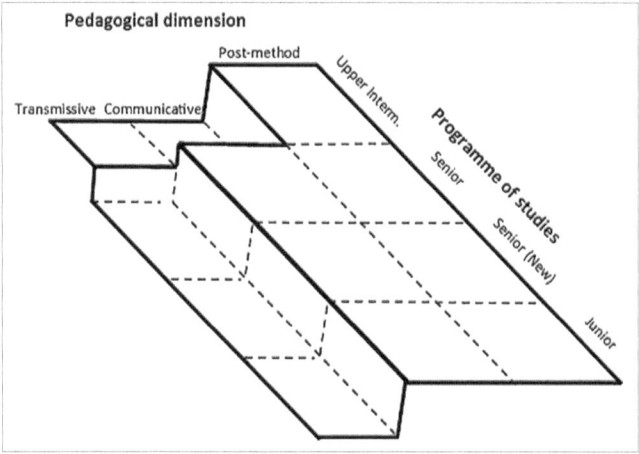

The salience of grammar in the learning activities was hinted by a cursory examination of the contents pages of the coursebooks. Despite organisational differences, the syllabus of every course invariably contained an outline of the grammar structures that were to be learnt in each lesson. Table 4.5 presents the grammar content of the first lesson of various coursebooks, as described in their contents page.

Table 4.5: Extracts of content pages

Book	Lesson 1 grammar content
Junior B	*be* (affirmative, negative, questions) Possessive adjectives
New Senior 1	Present Simple: *be* Possessive adjectives Possessive 's
Senior 2	Present Simple: routines, general truths, adverbs of frequency
Exam B1	Present simple & continuous Be used to Stative verbs

The insight regarding the prevalence of grammar were also quantitatively corroborated: More than a third of the activities in the published materials focused on developing grammar awareness and practicing grammar structures, which made grammar the most common type of activity in the book. The prevalence of grammar activities was considerably higher in the newly introduced *New Senior* materials, where 43% of the activities had an explicit grammar focus. This was due to the use of a dedicated grammar book, which contained extensive grammar practice activities. When I enquired a member of the senior management team about the use of a grammar book at this level, I was told that this reflected a deliberate pedagogical decision, prompted by teacher feedback, as well as what was described as the students' 'unsatisfactory performance' in this area.

In addition, many skills-focused activities, particularly in the Junior and Senior levels, tended to have an implicit grammar objective alongside the overt objective of practicing receptive and productive skills. For instance, many reading passages and listening texts were contrived to contextualise target structures, as is the case in this reading text from *Junior B*, which appears to have been contrived to showcase question and negative forms:

Learning Materials Example 1 – What is the matter, Rose? Do you want a sandwich or a biscuit? – No, I don't want a sandwich and I don't want a biscuit.

Similarly, a large number of activities that involved productive skills seemed designed to elicit grammar structures. For instance, a lesson in *Senior B* concluded with a 'speaking and writing' activity in which students had to describe how they celebrated their birthdays using a pre-defined set of verbs. This highly structured activity was likely intended to elicit utterances in the present simple tense, which had been the focus of the lesson in which the activity was embedded.

On the whole, grammar activities in the courseware tended to be transmissive. This was particularly the case in the Junior and Senior programmes. Extensive grammar presentations, in English, which were present in most lessons, hinted at the importance attached to explicit awareness of grammar structures, including the use of metalanguage. The following example is taken from *Senior B*, a course for 11–12 year old students at the A2 CEFR level.

> **Learning Materials Example 2**
> We use the present simple for things that happen again and again, or things that are always true.
>
> *He always wears his boots when he goes out.*
> *They love travelling.*
>
> We use the present continuous for something that is happening now, or something that is happening only for a short time.
>
> *I'm driving past the supermarket at the moment.*
> *I'm working on the other side of town this week.*

These grammar presentations were followed by limited production activities in a variety of formats, such as multiple-choice, multiple-matching, gap-and-cue, etc. The distribution of presentation and practice activities seemed to facilitate transmissive forms of instruction, such as Presentation-Practice-Production, or PPP (Harmer, 2015), whereas the predominance of limited production activities might suggest emphasis on accuracy and conformity.

This pattern of mainly transmissive grammar activities was less evident in the Upper intermediate programme. There was a moderate drop in the grammar activities in this level, and there was very little evidence of implicit grammatical objectives in skills-focused activities. Most importantly, grammar activities often aimed to engage the students' cognitive skills and to help them infer grammatical regularities on their own. Grammar reference sections containing detailed presentations of the target structures were available for self-reference at the end of the coursebooks, but they did not appear to form an integral part of the activity sequences. Controlled practice activities were in evidence in the Upper Intermediate courseware, but they tended to be few in number, and their format suggested that their main function was examination preparation, as they appeared to be modelled after common examination tasks. These activity formats seemed to generate more communicative affordances.

The grammar provision in the courseware was supplemented by large number of photocopiable worksheets that had been created by the staff of the language school over time. Many of these resources seemed to be more traditionally-oriented than exercises typically found in the courseware. Notes explaining the structure and use of tenses were particularly common. These notes were usually produced to supplement or replace grammar notes in the courseware, which were sometimes perceived to be insufficiently detailed or unclear. They were invariably written in English, and tended to use metalanguage extensively. Apart from grammar notes, a large number of worksheets consisted of grammar exercises, which often practiced form in decontextualised ways (e.g. by asking students to provide past forms of irregular verbs, or to transform sentences from Active to Passive voice). The use of such resources in all programmes of study, including the Upper Intermediate one, meant that transmissive affordances were always present in the system.

4.4 Affordances in the vocabulary activities

Vocabulary-focussed activities were the second most prevalent type of activities in the learning materials, and – much like grammar – they tended to generate strong transmissive affordances. These affordances connected to the expectation, implicit in the design of the materials, that the lexical items had to be explained or translated by the teacher, and then memorised and reproduced by the students. This was particularly the case in the two senior programmes of study, where the very large number of lexical items to be learnt necessitated time-efficient methods of teaching.

Figure 4.3: Affordance landscape (Vocabulary)

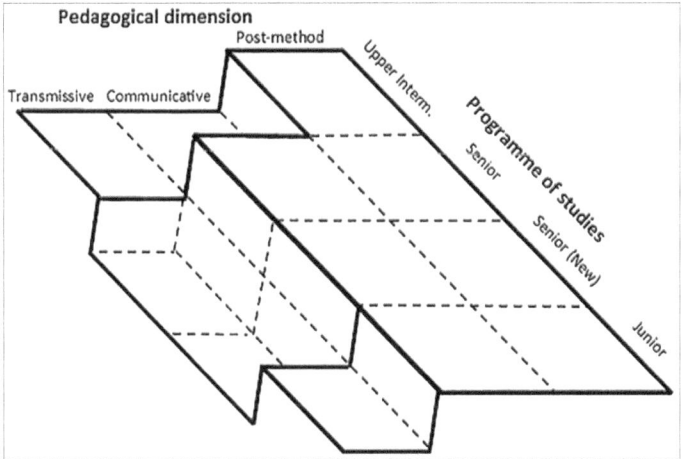

One thing that was striking in the learning materials was how extensive the provision for vocabulary learning was. In the published materials, approximately a quarter of the activities focused on vocabulary development (e.g., presentation of lexical sets and vocabulary practice tasks). This number was even higher in the Senior programmes, where the vocabulary activities accounted for nearly one in three activities in the materials. In addition, the materials in these courses were typified by a sometimes surprising lexical density. In *Senior B*, for example, learners routinely had to engage with, and usually memorise, more than 25 new lexical items (this was the case in approximately one third of the lessons). In two exceptional lessons, the number of new lexical items to be learnt was 45 and 49.

Most of the vocabulary activities in the coursebooks focussed on what I have termed *intentionally taught vocabulary*. The term refers to sets of six to eight lexical items (or fewer in the Junior programme) that the students were expected to learn. Normally, the criteria used for grouping the lexical items were semantic. For example, some lessons contained activities that focused on semantic sets like 'meat, fruit and vegetables' or 'farm animals'. Occasionally, lexical items were grouped according to formal criteria, such as a common base verb (e.g., 'multi-word verbs: turn') or other formal similarities (e.g., 'adjectives ending in -ed & -ing'), although these cases were less frequent. These activities were associated with transmissive affordances, as they typically involved memorisation.

A second type of vocabulary activities often encountered in the course books involved *incidentally taught vocabulary*. This category comprised various vocabulary activities that either preceded reading and listening activities, or followed them. Reading passages, for instance, were almost invariably accompanied by activities with a vocabulary-related goal, such as inferring the meaning of unknown lexis in a text. The lexical items that made up the incidentally taught vocabulary differed from the intentionally taught semantic sets, in that they were seemingly selected at random, as there were no perceptible semantic or formal connections among them. These activities sometimes required students to engage with the propositional content of passages in which the lexis was embedded in order to infer meaning, so they could generate communicative affordances.

Intentionally and incidentally taught vocabulary items were also encountered in a numerous practice activities, which made up more than half the content in some workbook units. A variety of exercise formats were encountered, including multiple-choice, matching and gap-and-cue exercises, as well as activities focusing on synonyms, antonyms and cognates. Most of these activities were associated with transmissive affordances, in that they involved the reproduction of memorised content.

The last vocabulary category, *independently occurring vocabulary*, made up the largest part of vocabulary in the syllabus. This term designated diverse lexical items that were extracted from the activities found in the coursebooks or, less frequently, the workbooks. The absence of any production activities associated with these items suggests that, from the course authors' perspective at least, this category was intended to form the students' receptive vocabulary (Melka, 1997). However, in the syllabus documents of the language school no distinction was made between productive and receptive vocabulary. In fact, an extensive corpus of exercises had been created over time by the school staff to address the lack of practice activities for this vocabulary category. The requirement that students memorised the meaning and form of these lexical items – in addition to those in the other two vocabulary categories – created additional transmissive affordances.

Apart from their coursebooks, students and teachers could draw on two main vocabulary resources: companions and monolingual wordlists, both of which also generated transmissive affordances. Companions were published booklets that accompanied coursebooks and functioned as vocabulary supplements. The main content of these booklets was a set of wordlists that paralleled the structure of the coursebooks. Each entry in the lists contained, at minimum, a keyword and its semantic equivalent in Modern Greek. This information was sometimes supplemented with definitions in English and phonological, grammatical and pragmatic information. These lists were supplemented by a small number of activities designed to reinforce recall of vocabulary items (e.g., gap-filling, multiple choice or matching activities) as well as study tips.

The use of companions was discouraged at the language school, because they were bilingual, and therefore incompatible with the English-only policy in place. However, several teachers and learners unofficially endorsed them as vocabulary aids, or tacitly accepted their use by students. Some hints about the popularity of these resources can be found in Kostoulas (2007), where it is argued that companions seemed to cater for idiosyncratic needs of the local educational setting. In that study, a key figure in a major publisher is quoted as saying that 'teachers say that they [i.e., companions] are useful to establish quick links with mother tongue' (p. 73). The head researcher of the local office of a major publisher elaborated on this view by suggesting that the bilingual lists in the companions:

> ...reflect the manner in which a reading text or dialogue is taught in class. The average teacher will 'comb' the text or even translate it in a manner very similar to the way words and phrases are entered in a companion. (Kostoulas, 2007, p. 73)

The view that companions were well suited to local teaching styles was corroborated by several teachers in the school. The following interview extract conveys a typical view:

Interview Extract 1
Achilleas: Tell me some more about the companion. Why is this so helpful?
Teacher: Because it has the words, Greek and English. It has pictures, it has examples, it has short activities, mostly vocabulary, to reinforce, to make kids practice the vocabulary they have in front of them.
Achilleas: Some people feel uncomfortable with the companion, though.
Teacher: I [do] not. I love the companion.

In other words, the structure of these resources seemed to be compatible with local teaching styles, and their content capitalised on the teachers' and learners' shared knowledge of Modern Greek.

Table 4.6: Format of monolingual vocabulary lists

	Headword	Grammar category	Definition	Cognates etc.	Examples
Junior	List not available				
Senior (New)	✓		English		Few
Senior	✓	✓	English	Few	
Upper Intermediate	✓	✓	English	Several	Several
Proficiency	List not available				

To replace the companions, the school management had commissioned the creation of monolingual wordlists. These resources replicated what were perceived as the more useful features of the companions, such as the structure and sequencing of their lists, but provided glosses and dictionary-like definitions in English, as opposed to Greek sematic equivalents. Table 4.6 summarises the content of these wordlists: each headword was followed by a definition in English, and typically a note identifying its grammatical category (e.g., 'noun', 'verb'). Cognates, synonyms and antonyms were sometimes listed, with increasing frequency as one progressed to the more advanced levels. In some cases, the use of the words in context was illustrated through example sentences. In the most recently produced wordlist, the one for the Senior (New) programme, the use of grammatical designations was dropped, and examples were introduced. According to the teacher who was responsible for the production of these wordlists, this reflected a conscious effort to accommodate to the needs of younger learners.

Both the companions and the wordlists were intended for memorisation. It was standard practice at the school to assign specific pages for study, and test recollection of their content in the next lesson (see also Section 6.2). In this sense, the availability of these resources seemed to generate transmissive affordances.

4.5 Affordances in the reading and writing activities

The activities connected to the development of reading and writing also created a set of transmissive affordances throughout the curriculum of the language school. These was more pronounced in the Junior and Senior programmes, but at the Upper Intermediate programme, they became weaker and were accompanied by a very strong set of communicative affordances (Figure 4.4). This shift was not always very evident in the prevalence of activities, but it was noticeable in their format.

Figure 4.4: Affordance landscape (Skills)

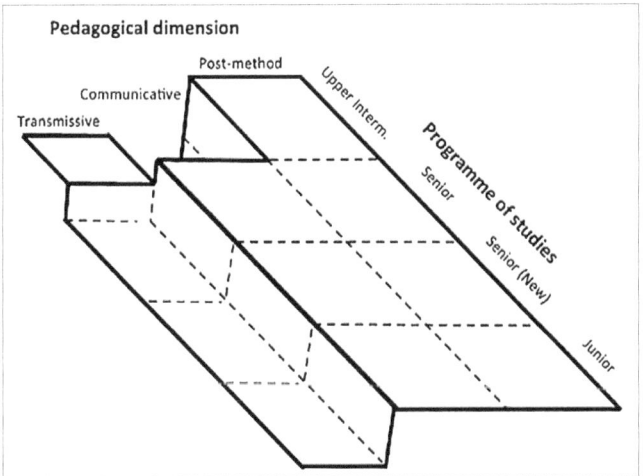

Writing activities made up about less than one tenth of the activities in the published materials, and were very unevenly distributed across the different programmes of studies. In the Junior and Senior programmes, there was a gradual decrease in their number. Starting from 12% of the activities in the Junior programme, they eventually fell to under 6% of the *Senior* materials. This trend seemed to coincide with a substantial increase of vocabulary activities in the Senior programmes, and a similar increase in the grammar activities in the Senior (New) programme. The drop in the prevalence of writing activities was yet another indication of the strongly transmissive outlook of that typified instruction at these levels: the expectation seemed to be that students should passively assimilate information, rather than produce language. However, this trend was reversed in the Upper Intermediate programme, where writing activities took up about one tenth of the learning materials, a trend associated with more transmissive affordances.

In addition to the cues provided by the quantitative change, there was also a remarkable qualitative differentiation associated with the transition from the Senior programme to the Upper Intermediate one. In the learning materials there were, broadly speaking, three distinct types of writing activities: form-focused writing, genre-based writing, and process-based writing activities.

Form-focused activities used writing as a means for consolidating previously taught language. Most commonly, they constituted the final parts of larger activity sequences that followed the Presentation-Practice-Production pattern. These were associated with strongly transmissive affordances, because of their connection to the transmissive teaching of grammar.

Genre-based activities were defined as writing tasks which entailed the production of text types with specified formal features (Tribble, 1996). For instance, learners might be tasked to produce stories, reports or email messages. Genre-based writing activities were often embedded within larger pedagogical sequences, which involved engaging with prototypical instantiations of the genre ('model' texts), and explicitly identifying salient features, such as its macro-structure and discourse markers. There was often a progression from more controlled modes of production to relatively freer ones. While not aiming to practice grammar, these activities also created transmissive affordances, as they involved memorising and reproducing formal properties of various genres.

Process-based writing activities were defined as activities that focussed on developing various writing sub-skills, such as planning, drafting, editing, etc. (cf. Seow, 2002). These activities tended to be embedded in larger pedagogical sequences that culminated in the production of a text, often through collaborative work (see also Section 6.4). The affordances with which these activities were associated were more communicative, as the activities aimed specifically at the development of writing competences, and because their implementation required task-oriented collaboration.

The distribution of writing activities across the curriculum suggested a shift in the affordances they created, with communicative affordances becoming more prominent in the higher levels. In the Junior programme, form-focused and genre-based activities were evenly balanced, but in the Senior programme, activities that aimed to practice previously taught grammar outnumbered activities that focused on the production of text types by 3:1. The Senior (New) programme contained a mix of all activity types, and there appeared to be a balance between activities that focussed on 'writing-to-learn' grammar and 'learning-to-write' extended discourse. Finally, in the Upper Intermediate programme, the writing strand of the syllabus seemed to be geared towards exam preparation, with some activities focussing on the formal features of specific text types, and the majority focusing on developing writing skills.

A similar pattern was evident in the prevalence and distribution of reading activities. Reading activities were especially common in the Junior programme, where they accounted for almost one fifth of the learning materials. According to the acting director of studies, this reflected the high importance attached to exposing learners to language before they were expected to produce it. In the Senior (New) programme, the prevalence of reading activities fell to almost half of that, and then it rose to about 14% in the next two programmes.

Although there was no clear quantitative evidence of increasing influence of the communicative approach in the more advanced levels, reading activities also evidenced some qualitative traces of such a shift. In terms of their format, the texts in the courseware were grouped into three categories, as follows.

Type A texts narrated events in the lives of recurring characters, and they were often displayed in cartoon or dialogue format. These texts were commonly used to contextualise grammar structures, or for practicing pronunciation skills through reading or acting, and were associated with transmissive affordances.

Type B texts were quasi-genuine passages that were thematically linked to the overarching topic of each unit. These passages were contrived for pedagogical purposes, but displayed formal features of real-life texts, such as websites, newspaper articles, agony aunt columns or song lyrics. Activities based on these texts were associated with mostly communicative affordances, as they required dealing with the propositional content of the text in ways that resembled real-life communication.

Type C texts functioned as models for subsequent writing tasks. Frequently-encountered genres included short personal narratives, email messages, film reviews, reports, and argumentative essays. The activities associated with these texts required engagement both with content and with form, and allowed scope for both transmissive and communicative affordances.

Figure 4.5: Distribution of text types across programmes

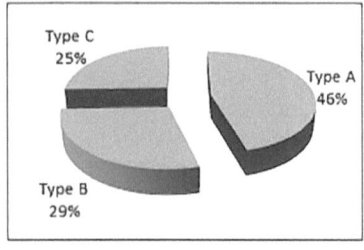
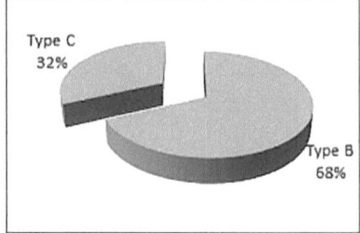

Junior & Senior   Upper Intermediate

As can be seen in Figure 4.5, the texts that were associated with transmissive affordances were especially prevalent in the early years of instruction, but in the Upper Intermediate programme of study, this pattern was disrupted in favour of more communicative affordances. In the Junior and Senior programmes (left), approximately half the texts were Type A, but this category was not present in the Upper Intermediate programme (right). Conversely, the number of Type B texts more than doubled in the Upper Intermediate programme, rising from less than one third to more than two thirds of the reading activities in their respective coursebooks. This shift towards more genuine-like reading materials seems to be associated with the content of reading tasks in the certification examinations.

4.6 Affordances in the listening and speaking activities

The affordances generated by the listening and speaking strands of the syllabus were quite similar to those generated by the activities connected to the written modality (Figure 4.4), except the trend towards more communicative affordances in the Upper intermediate level was more pronounced.

To clarify, I use the term listening activities to refer to activities in which students were expected to interact with the propositional content of aural discourse that had been recorded or was read out from a transcript. This definition excludes activities that mainly focused on pronunciation work, where recordings were used for demonstration, as well as 'read and listen' activities, during which students read a passage while a recorded version of the text was being played back. Using this definition, listening activities were found to be the least prevalent activity type in the learning materials, especially in the earlier programmes of study. Their frequency increased from one in forty in the Junior programme, to more than treble the number in the Upper Intermediate programme, but it never exceeded 8% of the total number of activities.

Depending on their pedagogical function, listening activities were divided into two categories. The first category, form-focused listening, comprised activities that were used in order to assist students in attaining other linguistic or pedagogical objectives, such as the consolidation of recently taught lexis or grammar. The following transcript, from the *Junior B* coursebook, is an example of such activities:

Learning Materials Example 3	
Shop assistant:	Good morning, Danny!
Danny:	Hello! I want some red paint and some paintbrushes, please.
Shop assistant:	How much paint do you want? And how many paintbrushes?
	[...]
Shop assistant:	Hello, Flora.
Danny:	Hello! I want some red paint, some green paint and some paintbrushes, please.
Shop assistant:	How much paint do you want? And how many paintbrushes?

Almost all the listening activities in the Junior and Senior programmes fell into this category, creating strong transmissive affordances.

The second type of listening activities involved an authentic engagement with texts, such as dialogues, interviews or radio broadcasts, for the purpose of skills development, and thus were associated with more communicative affordances. It should be noted that authenticity, in this sense, refers to the purpose of listening, rather than the provenance of the actual texts (Widdowson, 1979). For example, learners might listen to a simulated radio interview in ways that resembled real-life situations, such as for global comprehension, or for extracting specific information. The texts themselves were genuine-like, in that they conformed to the formal conventions of their genres, but appeared to have been written and recorded for pedagogical purposes. Most, but not all, of the activities in the Upper Intermediate programme belonged to this category. In the Upper Intermediate programme they also tended to be flanked by pre-listening activities intended to activate schematic knowledge, and post-listening tasks involving oral or written production (cf. Anderson & Lynch, 1988). Many activities in this category mirrored the format of the listening tasks encountered in certification exams, and explicit advice for engaging with such tasks was often at hand.

This pattern was even more pronounced in the speaking strand of the syllabus. Compared to listening, there was an even more marked increase in the provision for speaking practice in the Upper Intermediate programme, as well as a qualitative shift towards communicatively informed activities. Arguably, most activities in courseware involved some degree of oral production. However, for the purposes of this chapter, speaking activities have been defined as activities that were primarily designed to either (a) develop the students' ability to engage in the negotiation of meaning or (b) produce relatively extended pieces of discourse using the oral modality. The existence of visual or verbal signposting identifying the activity as a speaking task was also taken into account. Activities in which oral production was only incidental (e.g., reading out a passage for pronunciation practice, or reciting answers of a grammar exercise in order to receive feedback) were excluded from this definition.

Using these criteria, fewer than one in ten activities in the published materials focused on speaking, and nearly half of these were clustered at the Upper Intermediate programme. In fact, speaking activities accounted for almost 15% of the materials in the Upper Intermediate programme, which was almost twice or treble that encountered in the other programmes. As was the case with listening activities, this sharp increase seemed to be associated with the fact that the examinations for which the learners were preparing tested speaking directly. In addition to the distribution disparity, a qualitative differentiation was also observed in typical speaking activities, depending on the level of instruction. In the Junior programme, speaking tasks mostly took the form of short exchanges intended to practice previously taught vocabulary or grammar structures. The expected language output was repetitive, not dissimilar to language drills. Unlike drills, however, the recommended format of the speaking activities almost invariably involved pair-work rather than teacher-led interactions. In other words, it would appear that activities in the Junior programme blended traditional and communicative influences.

In the Senior programmes, speaking activities seemed to adhere to communicative methodological orthodoxy, but there were traces of underlying transmissive influences as well. Most of the activities in these levels were designed for pair- or group-work. Others were flexibly structured so that collaborative work was possible but not required. A small number of activities were suitable for teacher-led discussions. In the syllabus outlines, these activities were usually described with reference to their topics (e.g. 'Talk […] about speaking foreign languages', 'Talk […] about what you have done'), but they were often implicitly linked to specific language structures. In short, the speaking activities struck a balance between conformity to communicative methodological orthodoxy, and accommodating to the transmissive practices with which teachers and learners were most comfortable.

This balance between transmissive and communicative influences was disrupted in the Upper Intermediate programme. Approximately a third of the speaking activities were embedded in longer sequences, as lead-in or follow-up to other tasks. These activities were flexibly designed, so that they could be used as either prompts for teacher-led discussions or as scaffolding for pair- and group-work. The remaining activities, however, were clustered in sequences that seemed to draw on Task-Based-Learning methodology (Willis & Willis, 2007). Student collaboration was built in the task structure, and it was explicitly signalled through rubrics and visual cues. Extensive support was provided through recordings of model conversations, word lists containing appropriate discourse markers, and explicit instructions on discourse management. There are multiple clues in the activities indicating that this communicative turn was associated with examination preparation, including the formal similarity of speaking activities with common examination tasks, and the provision of explicitly signposted examination strategies.

4.7 Putting it all together

The description of the learning materials in the previous section suggests that the learning materials at the language school tended to privilege transmissive or communicative pedagogical practices. To connect this with the discussion of the state space outlined in Chapter 3 (Figure 3.1), the affordances, or action possibilities, associated with these materials seemed to align to with the transmissive and communicative positions of the pedagogical dimension.

The distribution of these activities, and by extension the distribution of the affordances, was not uniform across the curriculum of the language school. Rather, the affordance landscapes varied dynamically along three dimensions. One dimension, which might be called *spatial* variation, concerned the uneven distribution of affordances in different syllabus strands. The majority of the activities in the learning materials were associated with the transmissive position. This was most obviously the case with the activities involving the presentation and controlled practice of grammatical forms, as well as most of the vocabulary activities, which involved memorisation. It was also the case with reading and writing activities which seemed designed to illustrate and use grammatical points or text genres. A small number of speaking activities that aimed at developing grammatical accuracy also fall into this category.

Table 4.7: Spatial variation of affordances

Transmissive affordances	Communicative affordances
Grammar presentation Grammar practice Vocabulary Type A reading passages Some Type C reading passages Form-focused writing Genre-based writing Some speaking activities	Grammar (inductive discovery) Type B reading passages Some Type C reading passages Process-based writing Most speaking activities Listening activities

A smaller, but still substantial number of activities generated communicative affordances. These included some inductive grammar activities, in which learners inferred grammatical rules from textual examples, as well as speaking, listening and reading activities, in which learners were expected to engage in meaningful communication. Lastly, process-writing activities were also included in this category, both because of the fact that they aimed at the production or personally relevant texts, and because they usually took the form of collaborative tasks.

Table 4.8: Phase variation of affordances

	Level	Transmissive affordances	Communicative affordances
Transmissive Phase	Junior A	✓	
	Junior B	✓	
	A Senior (New)	✓	(✓)
	B Senior	✓	
	C Senior	✓	
Balanced Phase	D Senior	(✓)	✓
	E Senior	(✓)	✓
	Proficiency 1	(✓)	✓
	Proficiency 2	(✓)	✓

A second aspect of variation, or *phase* variation, referred to the way in which affordances changed across levels of instruction. Broadly speaking, instruction at the language school seemed to be divided into two phases with different affordance patterns (Table 4.8). During the first, or Transmissive, phase, which comprised the Junior, Senior and Senior (New) programmes, transmissive affordances appeared to be predominant. This was evident both in the high salience of grammar and vocabulary activities encountered in these programmes, and in the fact that reading and writing activities very often appeared to have been contrived to respectively showcase and elicit language structures.

This pattern, however, was somewhat abruptly discontinued when learners reached the Upper Intermediate programme (D and E Senior). During this second phase, which might be termed Balanced, communicative affordances seemed to be more frequent, even though transmissive ones did not disappear altogether. Specifically, there was a marked increase in the number of activities practicing the oral skills, as well as a qualitative shift in the format of reading and writing activities, which tended to mirror the specifications of communicative examinations. Alongside this, there was also a change in the format of grammar activities, and inductive discovery activities became more frequent.

Table 4.9: Diachronic variation of affordances

Year / Materials	Transmissive affordances	Communicative affordances
2003 / Senior	✓	
2005 / Junior	✓	
2010 / New Senior	✓	(✓)
2010 / Exam	(✓)	✓

A final way in which the distribution of affordances varied was *diachronic*, and related to the way in which the language evolved over time. Evidence of this diachronic variation could only be observed indirectly, by examining the affordances created by different materials and relating them to the time in which they had been introduced (Table 4.9). The earliest set of materials in use, namely the Senior series, had been introduced in 2003. These coursebooks comprised activities that were mostly associated with transmissive affordances. These included activities for the transmissive teaching of grammar, an extremely strong emphasis on vocabulary learning (involving, in one case, the memorisation of 49 new lexical items in a single lesson), and reading, listening, speaking and writing activities with a clearly discernible grammatical focus. A similar affordance pattern was observed in the materials used in the Junior programme, which had been introduced in 2005.

The materials introduced in 2010, however, seemed to have a different affordance pattern. The *New Senior* coursebook and activity book generated more communicative affordances than the set of materials that they were replacing.

For example, there was a stronger emphasis on process writing, and the newly designed vocabulary lists de-emphasised information about grammatical form (e.g., grammatical categories), which was replaced by information about usage (e.g., examples). On the other hand, the perception that the new coursebook did not provide adequate grammar coverage led to the introduction of a dedicated grammar book, and a marked increase in the prevalence of transmissive affordances in the New Senior programme. This was not the case with the other course that was introduced in 2010, *Exam*, as the activities that it comprised tended to generate more communicative than transmissive affordances.

In summary, it seems that the affordance patterns generated by the learning materials were highly dynamic and complex. Various combinations of affordances were prevalent in different aspects of the school's curriculum, and these combinations varied in different phases of instruction and over time. To be clear, though, the effect of these affordances was not deterministic. Teachers and learners could re-interpret activities in ways that were closer to their own pedagogical preferences, and indeed did so, as will be seen in Chapter 6. The ways in which these affordances did influence practice, however, was by privileging specific pedagogical possibilities, and making it more likely that specific teaching and learning activities took place.

In Chapter 3, I defined a state space, which encompassed all the states potentially available to a system like a language school. I also noted that some regions, or constraining structures, in the state space were made more likely because of top-down constraints generated by established practices, teacher training, etc. In this chapter, I extended this discussion by suggesting that the activity of the system may also be channelled to specific regions, because of the affordances created by learning materials. The discussion in this chapter only focussed on those affordances associated with the pedagogical dimension of the state space. This was done for illustrative purposes, taking into account that pedagogical variations evidenced the most variation in my data, whereas the linguistic and political affordances seemed to be more uniform. A full description of a system, be that a language school, the education system of a country, or a single learner, would of course involve discussion of all the affordances present at a particular point in time, as well as a discussion of their dynamical variation.

So far, our description of a language school as a complex system has included the following elements: A state space of potential teaching and learning states, a set of constraints and affordances that make some states likelier than others, and a set of observations describing how the affordances are in flux. What is missing, to conclude the description of its structure, is a discussion of the forces that animate this system, and this will be the focus of the next chapter.

5 Driving activity in the system

Complex systems, sometimes also called complex *adaptive* systems, are typified by the fact that they are in constant flux. Even when they appear unchanging, this apparent stability belies the constant interaction between their constituents and entities that surround them. All this activity is driven and sustained by forces that emerge within the system – the technical name of these forces is *intentionalities* (Stelma, 2014). In this chapter, I examine the intentionalities that animated the language school, show how they came into existence, and how they interacted with the affordances, in order to pave the way for the description of how the system actually acted, in Chapter 6. This description is grounded on the ELT literature, with particular emphasis to the Greek context, as well as interview and questionnaire data from the language school taken from Kostoulas (2015a).

I begin this chapter by defining what intentionalities are and by describing their properties (Section 5.1). This will provide us with the conceptual tools and the technical vocabulary for the description that ensues. Next, I trace five intentionalities that were evident in the language school, which I connect to the affordances that were identified in Chapter 4. The first of these intentionalities, *Certification*, refers to a drive to certify language proficiency (Section 5.2). In Section 5.3, I discuss a concern, particularly among students, to integrate into transnational discourse communities, which I have called the *International Integration* intentionality. The next intentionality, *Cultural Awareness*, is a desire, especially among teachers, to impart cultural values associated with the Anglophone West (Section 5.4). Next, I discuss an intentionality that I termed *Competition*, and which refers to concerns about the way the language school related to state education (Section 5.5). The last intentionality that I will look into, in Section 5.6, is *Protectionism*, an unstated agenda of protecting the vested interests of local ELT. After presenting these intentionalities separately, in Section 5.7, I define their synthesis as a 'dynamic of intentions', and hint at its likely effect on the activity of the language school.

5.1 Defining intentionality

Intentionality, the key theoretical construct around which this chapter is structured, is defined as a self-organised collective driver of a system's activity, which leads to a specific outcome. This conceptualisation is consistent with the 'folk-psychological' definition of intentionality as 'purpose' or 'intention' (Stelma, 2014), which has informed research into educational technology (Young et al.,

2002), classroom interaction (Barab et al. 1999; Papadopoulou, 2011; Stelma, 2014), higher education (Stelma, 2011; Stelma & Fay, 2014), as well as ELT (Kostoulas & Stelma, 2017; Stelma, 2014; Stelma et al., 2015). The key properties of intentionality, which will be discussed in the following paragraphs, are that it is *collective, nested, emergent* and *generative*.

5.1.1 Intentionality is collective

While some studies of intentionality have focused on the intentional behaviour of individuals (e.g., Stelma, 2011; Stelma & Fay, 2014; Young et al., 2002), intentionality need not be understood in an individualistic sense. For instance, in Stelma (2013) the construct is brought to bear on the interpretation of classroom discourse produced by pairs of learners. In addition, Young et al. (2002) argue for a hierarchical conceptualisation of intentionality. In their model, lower-order intentionalities, such as those associated with walking to a polling station and casting a vote, might be seen as a component of a higher-order intentionality embodied in a political agenda. Even though Young et al. do not make the point explicitly, such changes in the scale at which activity is studied normally involve an analytical shift from individuals to increasingly broader collective entities. The perspective that has been adopted in this book, which aims to describe the activity of a language school, is based on a collective interpretation of intentionality. To illustrate, in Section 5.2 I argue that one of the intentionalities driving activity at the language school was providing learners with language certificates. Such references are meant as pertaining to the language school as a collective entity, even if they were not representative of the individualistic intentionalities of each and every agent in the system.

5.1.2 Intentionality is nested

The relation between intentional activity and the environment in which it develops is central to ecological psychology. In Young et al.'s (2002) ecological model, this relation is described in deterministic terms. Activity is described as being embedded in a cascade of constraints, ranging from the constraints posed by the logically feasible world, through additional levels of constraints associated with physical possibility, the laws of nature, the ecology in which the activity develops, and the actual world, to the constraints created by the socio-cultural reality. At each level of this hierarchy, the new set of constraints limits the degrees of freedom available for intentional activity to develop.

Young et al.'s model, though intuitively applicable to the natural world, has not withstood empirical scrutiny when applied to the social sciences. Accommodating agency has been one particular concern, as it has been convincingly argued that 'individual agency operates within and through social structures, but is not necessarily subjugated to them' (Hopwood, 2010, p. 832). Stelma (2011) eschews this difficulty by suggesting a heterarchical conceptualisation of context. In his model, Young et al.'s cascading levels of constraints are replaced by overlapping sets of 'shaping influences', akin to the interpenetrating systems described by Byrne & Callaghan (2014). In a description of the increasingly intentional way in which doctoral students made use of information technology, Stelma (2011) points to the ways in which doctoral projects were influenced, but not determined, by shaping influences including the expectations of supervisory teams, their School and University, national and international entities, as well as the resources that were made available by all of the above.

Shaping influences, which might be loosely defined as the effects exercised by ideological beliefs and resources, differ in terms of the power they exert and in their topological distribution. Based on CST, one might hypothesise that proximate entities exert more powerful influences than distally placed ones (Davis & Sumara, 2006). Such a hypothesis stands in contrast with deterministic narratives of ELT, which foreground the role of powerful influences stemming from the Anglophone West, while downplaying local activity (e.g., Phillipson, 1992). To better understand the role of shaping influences, in the discussion that follows, I place particular emphasis on what I call their *locus of origin*. Depending on the locus of origin, shaping influences are classified in three groups: (a) intrinsic shaping influences, stemming from within the system; (b) proximal shaping influences, i.e., influences that originate in the system's local context, such as the state schools that the students attended, their families, the training programmes in which the staff had participated etc.; and (c) distal shaping influences, i.e. influences that emanated from the global context, such as the globally marketed professional literature, or the effect of international examinations.

5.1.3 Intentionality is emergent

Although intentionality is sometimes used in the literature to describe 'volitional, conscious and explicit behaviour that aims at the fulfilment of a clear and recognisable target' (Papadopoulou, 2011, p. 579), explicit awareness of purpose is not an essential feature of intentionality, as understood here. Rather, intentionality is viewed as a phenomenon that emerges from the 'intentional dynamics' (Young et al., 2002, p. 50) that develop among multiple shaping influences. This conceptualisation is in line with Barab et al. (1999), who argue that:

> We further argue that the ecologized, or self-organization, model (relational ontology) establishes that (under the appropriate conditions) the particles (learners), in effect,

'want' to or strive opportunistically to order themselves once the intention has been properly initialized (p. 350).

For instance, in Section 5.4, I point out that Anglophile attitudes among the language school staff, a sense of cultural custodianship that English language teachers seem to share, and Anglocentric references in the learning materials all contribute towards the emergence of the Cultural Awareness intentionality. It should not be assumed that all agents in the system would be consciously aware of this intentionality, or that there was a deliberate alignment to it. However, the absence of explicit awareness at the individual level does not preclude the operation of the intentionality at the collective level.

5.1.4 Intentionality is generative

Lastly, intentionality is not viewed as a mere product of its environment, as might be suggested by Young et al.'s (2002) model of cascading constraints. Rather, it carries the potential to affect changes in the environment, create possibilities that would otherwise not be present, and thus add degrees of freedom to the system (Stelma, 2011). This potential was demonstrated by Stelma and Fay (2014), who described how the developing intentionality of novice researchers impacted the quality of their learning experience. In order to illustrate this property, in the discussion that follows, each of the five intentionalities is associated with a specific pedagogical effect, or *orientation*. The pedagogical orientation refers to whether this intentionality was directed towards the transmissive, communicative or critical pedagogy position in the state space (Chapter 3). This will allow us to connect the description of intentionalities to the discussion of affordances in Chapter 4. Similar orientations can also be defined in connection to the linguistic and pedagogical dimensions, but this is not done here, as my intention is to demonstrate how intentionality operates, rather than provide a comprehensive description of all their potential effects.

Having made these general comments about intentionality, I will now proceed with some illustrative examples of intentionalities that were present in the language school. My intentions, in doing so, are three-fold. First, I want to exemplify how intentionalities emerged from within the system. To that end, in each of the sections that follow, there is a discussion of the shaping influences that came together to bring the respective intentionality into existence. Secondly, I want to show how intentionalities pushed the system towards specific directions. To do this, I discuss the orientation of each intentionality, and its potential generative effects. Thirdly, I use the description of intentionalities that were present in the school in order to showcase specific theoretical aspects of the intentionality construct. Therefore, each section that follows will conclude with some theoretical remarks prompted by the empirical description.

5.2 Proving proficiency

One of the most salient intentionalities evident in the language school was what could be called *certification*, the expectation that language learning should lead to formal acknowledgement in the form of a certificate. The certification intentionality appeared to emerge from the interaction between beliefs about language learning and assessment and affordances generated by the certification options available. The certification intentionality seemed to be associated with the standard language ideology and a neutral political orientation; also, somewhat unexpectedly, given the prevalence of transmissive pedagogy in the language school (see Chapter 6), it had a strongly communicative pedagogical orientation. The dissonance between the orientation of this intentionality and actual practice shows that intentionalities do not operate deterministically.

5.2.1 What was so important about certification?

The certification intentionality seemed to stem from proximal and intrinsic shaping influences, i.e. shaping influences that had their locus of origin in the local context of the school as well as within the boundaries of the system.

Starting with the proximal influences, descriptions of ELT in Greece make frequent reference to a credentialist teaching and learning ethos (e.g., Karavas-Doukas, 1995; Mattheoudakis, 2007; Sifakis, 2009). Angouri, Mattheoudakis and Zigrika (2010) argue that language learners in Greece tend to be primarily motivated by a desire to obtain language certificates, a consideration which often outweighs motivation to learn the language as such. In their paper, they cite a parent as having explicitly framed her expectations as follows: 'I told my daughter I expect you to get (names certificate); you can learn the language later' (p. 192). The perceived value of language certificates is associated with the widespread belief that they might facilitate access to the job market, including coveted posts in the civil service and teaching posts in language education (Papaefthymiou-Lytra, 2012). A common limitation of these descriptions is that they are usually undertheorised, or alternatively they seem to understand pressure for certification as a unidirectional deterministic influence, which often ignores the complex ways in which societal expectations interact with both agency and other contextual influences. However, they are helpful in contextualising the shaping influences which were intrinsic to the language school.

Looking at the school and the intrinsic shaping influences that developed inside it, I conducted two questionnaire surveys in which I asked learners about their views concerning learning and certification. In the first survey, younger learners whose linguistic proficiency in English was at the A2-B1 level of the CEFR were asked to respond to Likert-type items which probed their attitudes towards exams. Even though the learners were not enrolled in an examination preparation course, and most them were not expected to take a certification exam in the next two or three years, almost all of them provided strongly positive answers to the question 'Are examinations such as [names of well-known examinations] important to you?' In fact, nearly nine out of every ten respondents agreed, with varying degrees of conviction, that obtaining a language certificate was even more important than a good command of the language.

These findings were confirmed by the second questionnaire, in which learners enrolled in the Upper Intermediate programme were asked to answer a number of open-ended questions. In one of these questions, participants were asked to provide advice to an imaginary friend who wanted to stop learning English before obtaining a certificate. Specifically, the imaginary student in the scenario asked:

> I've learnt some English (up to D' Class) but I don't think it's interesting any more. I already know enough, so I'd like to stop now and start a new language, like French or German. What do you think?

In the students' responses, the view was unanimously expressed that the imaginary student should not interrupt her studies at this stage, and many respondents explicitly recommended that she should acquire some form of certification before quitting. Typical responses included the following:

Student Quote 1
In my view, you should at least take the first certificate as it is necessary for the job.

Student Quote 2
I would recommend that you take at least the ECCE certificate (for job reasons) and then if you would like to stop, stop.

Student Quote 3
I don't disagree with the choice of starting a new language, but in my opinion it is better to continue learning English at least up to Lower [B2] level.

Student Quote 4
She could continue for a couple of years her lessons, in order to take [illegible] her certificate. Then her English will be recognized.

These responses suggest that in the students' perception at least, the value attached to holding a certificate outweighed any benefits of being able to communicate multilingually.

Elsewhere in questionnaires, students discussed the importance of certification, with reference to the instrumental goal of improving their academic and employment prospects. While some students acknowledged that '[it] is more important to know and speak English rather [than] to have the papers and not use [the language]', the prevalent attitude seemed to be encapsulated in the following statement: 'It's more important to speak and write English but without a diploma it's like you don't know anything'. In a commonly expressed view, 'you need these papers to find a good job' because a certificate 'improves [proves?] that you are good in English'. In addition, another student stated that having such a certificate means 'you can find easier job and you can also study in England in very well [i.e., good] universities.'

Similar views were expressed by the teachers in the language school in a series of interviews. All of them agreed that obtaining some form of certification constituted a priority for students, although they sometimes expressed personal ambivalence about how appropriate this goal was. The importance of certification was explained by some teachers as a product of intense societal pressure. In the words of one teacher:

Interview Extract 2

Teacher:	I have students who don't want to come to the classes, who cry, who say they 'I don't like English but I have to do it in order to find a job and in order to satisfy my parents.' And I think that they don't learn English just for their own sake. They learn it in order to take a degree, to gain a degree. To hold a degree.
Achilleas:	How many students do you think that is? I mean do you think that is a small percentage of students? Many students? How you describe that?
Teacher:	From my personal experience, I would say a large amount of students. Yeah. They are. And I don't know if it has to do with the place where I work, but I if I go back to my school years, I think that a lot of students didn't want to learn English-
Achilleas:	Didn't really want to learn English?
Teacher:	They were forced to.

These views were echoed by another teacher, who remarked that:

Interview Extract 3
For some learners there is also the element of fun, the intrinsic motivation of learning a new language. This was mainly true in the smaller levels, where activities tend to be better suited to the interests of young learners. But overall, the main reason students come to the school is because their parents expect a 'lower' or a 'proficiency' [B2 and C1 level certificates].

The same teacher went on to explain that parents and students seemed to care more about obtaining a certificate, rather than about acquiring the knowledge that it certified. She supported her belief by pointing out that parents who came to consult with her were usually very keen to know when their children might take the exams, but 'they never ask about what words they [i.e., the students] know, [or] what they can do with the language'.

In addition to societal pressure, some teachers hinted at other considerations that contributed to the development of the certification intentionality. For instance, one teacher claimed that making certification a goal of language instruction enhanced the students' motivation. She went on to explain that 'not all the children in the class are always interested in learning more grammar or more vocabulary', but certification offered a real goal which was important to them. In this sense, certification seemed to address the lack of motivation inherent in what would otherwise be a Teaching English for No Obvious Reason (TENOR) situation (Abbot, 1981; see also Xanthakou, 2005, for a discussion of TENOR in the Greek context). Another teacher suggested that language proficiency examinations helped to maintain standards of professional excellence both within the school and in Greek ELT more generally. In the view of this particular teacher, encouraging learners to prove their linguistic proficiency was beneficial because it to provided external validation of the 'excellent quality education' that the language school offered. A third teacher also remarked that 'even though standards [in Greek ELT] are declining', certification exams are still important in motivating learners to excel, and added that students 'are certainly not harmed' by being encouraged to study harder or by receiving independent confirmation of their linguistic proficiency.

These views were also confirmed by the acting Director of Studies at the language school, who suggested that encouraging learners to obtain outside certification was beneficial to the school, in addition to having 'obvious advantages to the students themselves'. It was emphatically pointed out that there was no direct financial benefit for the school as such, and that any pecuniary incentives provided by some examination providers were too trivial to merit consideration. However, it was felt that the school's success rates, which were consistently above average and compared very favourably against commercial competitors, served to enhance the reputation of the language school, and hence (it is to be assumed) its long-term commercial viability.

Although the views expressed above suggest a strong acceptance and even ownership of the certification intentionality, this was not always uncritically accepted. For example, in one of the interviews, a teacher voiced her concern that the conflation of learning the language and obtaining a certificate was 'the most important problem facing ELT in Greece', and suggested that societal expectations and commercial interests were equally to blame for this problematic situation. Another teacher lamented the fact that the priority attached to testing, and especially formal examinations, compressed learning time. She pointed out what she described as a regrettable trend, where 'some *frontistíria* [private language centres] are even advertising that one can obtain a certificate even after five years of studies', as opposed to six or seven years, which she felt were necessary for a young learner to reach the B2 level of linguistic proficiency. This resulted in what she described as 'shallow, hollow' knowledge of the language. Similar sentiments were echoed by several teachers, who felt that the combination of pressure to reach the examinations in less time, coupled with mechanical teaching to the test practices were responsible for the poor academic results and lack of lasting learning outcomes that typified Greek ELT as a whole.

To briefly recap, the certification intentionality was associated with a number of shaping influences. Predominant among them were widespread societal expectations about the value of linguistic certificates. These proximal shaping influences were reinforced by beliefs held by the learners and teachers at the language school, some of which were grounded on pedagogical considerations and some of which connected to the status and commercial viability of the school.

5.2.2 What certification options were available?

In addition to the beliefs and expectations outlined above, the certification intentionality was also shaped by the specifications of the examinations that were available to learners in the language school. This was done in two ways. Firstly, the examinations exerted a direct influence on the orientation of the intentionality by legitimising specific language norms and task formats. Secondly, they influenced the intentionality indirectly through the effect they had on learning materials. As seen in Chapter 4, many learning materials contained activities that mirrored the format of examination tasks, and thus created specific affordances for teaching and learning. The combined effect of these influences was that certification had a normative linguistic orientation, associated with native speaker norms and the standard language ideology, as well as a strongly communicative pedagogical orientation. To better understand how these orientations came into being, in this section we will take a closer look into the various certification options. This description draws on open-ended questionnaire administered to learners, teacher interviews, documentation provided by the management of the school, and publicly available information.

There seemed to be a shared expectation among teachers, students and their parents that at some point during their language studies, the majority of students would obtain a B2-level certificate, which was colloquially described as a 'lower'. Although this was not explicitly acknowledged in the curriculum and marketing documents of the language school, the Upper Intermediate programme was designed to prepare students to receive such a certificate, and it culminated in an extended period of exam preparation. Following successful examination at the B2 level, many students would then enrol in the Proficiency programme, which prepared them to sit a C2-level examination. A C2-level certificate (colloquially called a 'Proficiency') indicated the ability to use English 'with the degree of precision, appropriateness and ease that typifies [...] highly successful learners' (Council of Europe, 2001, p. 36). More importantly, it conferred upon its holder the right to apply for an English Language Teacher's licence after they had completed secondary education (and military obligations, in the case of young men). Although precise information about student enrolment and progression was commercially sensitive, and therefore cannot be published, virtually all (>90%) the students who enrolled in the General English programmes at the language school would eventually take a B2-level examination, and the number of students who went on to take C2-level exams was approximately a third of that.

It was standard practice for students who finished a language course to register for at least two, and sometimes three, examinations of the same level in order to maximise their prospects of success. The Cambridge ESOL examination suite was the set of language certificates preferred by the language school. These tests were designed by Cambridge ESOL, an examination board affiliated to the University of Cambridge. The most popular certificates were the *First Certificate in English (FCE)* and the *Certificate of Proficiency in English (CPE)*, which corresponded to the B2 and C2 levels respectively.[1] These communicatively-oriented examinations consisted of four components ('papers') that directly tested the Reading, Writing, Listening and Speaking skills, as well as a paper that tested grammar and vocabulary (Use of English).

The Cambridge certificates were considered the most prestigious options available to students. According to some teachers, the high standards associated with these examinations encouraged students to excel. It was also felt that success rates at these examinations were a particularly good indicator of quality of prepara-

---

1 These examinations have been since rebranded as *Cambridge English: First*, and *Cambridge English: Proficiency*. The terminology used in here will reflects usage at the time of the study. A special version of the *Cambridge English: First* is now available for school-age candidates, but did not exist at the time when data was collected.

tion. Among the learners, the prestige of these certificates was undisputed. One prospective candidate described them as entailing 'good prospects of finding a job worldwide', and another claimed that these certificates were 'recognised everywhere'. It was also suggested that this examination was especially suitable 'for students who are very good at English'. However, many learners felt uneasy about the examinations through which the certificates were conferred: some participants felt that the time-intensive examination (in excess of four hours) was too exacting and many believed that the examination was disproportionately hard, compared to alternatives.

The main alternative to the Cambridge ESOL certificates were the certificates offered by the University of Michigan English Language Institute (UMI/ELI), namely the *Examination of Communicative Competence in English (ECCE)* for the B2 level, and the *Examination of Communicative Proficiency in English (ECPE)* for the C2 level. These examinations consisted of Writing, Listening and Speaking components, as well as a component blending Grammar, Vocabulary and Reading (plus several cloze tasks at the ECPE test). A distinctive feature of these tests was the preponderance of multiple-choice items, which were extensively used to test grammatical competence, vocabulary range and the receptive skills. The tests that led to these certificates were generally regarded as being easy compared to those designed by Cambridge ESOL, a fact that made them popular among learners. However, concerns were voiced among some teachers that by encouraging students to apply for such tests entailed 'diluting standards'.

A third certification option was offered by the London Tests of English (LTE, later rebranded as the Pearson Tests of English), which were designed by Edexcel, and administered in Greece by a local educational agency called London Exams Hellas, in conjunction with the Pan-Hellenic [National Greek] Association of Language Schools Owners. The examinations conferring B2 and C2-level certificates were known as *LTE Level 3* and *LTE Level 5* respectively, although they were more commonly referred to as the 'Edexcel Lower' and the 'Edexcel Proficiency'. These examinations comprised a succession of interlinked communicative tasks that simulated a real-life scenario (e.g., taking notes while listening to an answering machine, and then using them in order to draft a letter). This examination format was quite popular with some learners, but it seemed to compromise the 'face validity' (Alderson, Clapham & Wall, 1995) of the examination: according to some teachers and students it was not a 'proper' test since it did not explicitly test vocabulary or grammar.

A less popular certification option was the State Certificate of Language Proficiency («Κρατικό Πιστοποιητικό Γλωσσομάθειας», usually shortened to *Kratikó* or KPg), which was designed by the University of Athens Research Centre for English Language and administered by the Hellenic Ministry of National Education and Religions (currently the Hellenic Ministry of Education, Research and Religions). When the study took place, there was a certificate on offer for the B2 level, but not for C2. The examination consisted of four components, testing Reading and Linguistic Awareness, Writing, Listening Comprehension, and Speaking. A distinctive feature of the examination was that it placed emphasis on the candidates' mediation skills, which were operationally defined as the ability to summarise information in Greek and present it to an English-using audience using appropriate register and levels of detail (Karavas, 2009).

The KPg certificate was by far the least popular option among teachers and students at the language school. The KPg examinations appear to have developed a reputation of unprofessional and bureaucratic administration procedures, as inferred by anecdotal evidence that several teachers shared with me. Teachers also criticised the paucity of feedback on examination results, and what they perceived to be inconsistent grading patterns. In addition, a belief seemed to exist that local language certificates lacked international validity, or (in the words of a student) that 'it is not so important, because it is not accepted by foreign countries'. It appears, in other words, that this examination also suffered face validity challenges, although these seemed to be primarily related to beliefs about the efficacy of the entities associated with it, rather than with the format of the examination as such.

Certification options that have not been listed above, such as City and Guilds, the Test of Interactive English (TIE), the Test of English as a Foreign Language (TOEFL) and the International English Language Testing System (IELTS) tended to fill niche functions and their market share was relatively low at the time. As a result, the language school did not regularly prepare students for these examinations.

A feature common to these examinations was the casual way in which they promoted native speaker norms, which seems to relate to the Standard Language Ideology (Section 3.1.1), and uncritical theorisations of ELT (Section 3.3.1). Two of the major examination boards, Cambridge ESOL and University of Michigan English Language Institute (UMI/ELI) were based in the United Kingdom and the USA respectively, and these appeared to enjoy the best reputation and greatest prominence. A third major examination provider, London Exams Hellas, was part of a Greek conglomerate, but relied on a UK examination board for their testing materials. In addition to the physical location of the boards, the Anglocentric influence of these examinations was evident in the fact that they tended to prioritise Standard Language varieties. The Cambridge ESOL and UMI/ELI examinations almost exclusively used British and American English in their listening materials (Kanellou, 2012). Moreover, Cambridge ESOL documentation indicated that a consistent use of American or British spelling was desirable (University of Cambridge Local Examinations Syndicate, 2007) and the UMI/ELI examinations were explicitly self-described as examinations of Standard American English (Irvine-Nikiaris, 2009). Lastly, most examination boards made use of Anglocentric branding. This was particularly evident in the case of London Exams Hellas, who evoked associations with the UK in their name, and prominently marketed their association with Edexcel, their international partners.

The State Certificate of Language Proficiency (KPg) claimed to be exceptional in this regard, although the validity of this claim does not hold up to critical scrutiny. The examination, which was designed and administered by a Greek examination board, was explicitly self-described as having a 'pluricentric' outlook:

> As far as the KPG exams in English are concerned, it should be noted that English is viewed as a pluricentric language and the KPG exams deal with World Englishes. As a result, oral and written texts are not exclusively in British or American English but may be, for example, in Australian or Canadian English. (University of Athens Research Centre for English Language, 2012, p. 3)

However, 'pluricentrism' was idiosyncratically defined with reference to non-prototypical native-speaker settings, rather than Outer or Expanding Circle ones. Somewhat surprisingly in light of their professed pluricentric outlook, the University of Athens Research Centre for English Language, which designs the KPg examination, has advertised vacancies for staff with descriptions such as the following:

> Άτομα που έχουν την αγγλική ως μητρική γλώσσα με κάποιου τύπου θεατρική αγωγή (για να παίξουν ρόλους σε επικοινωνιακές περιστάσεις που θα ηχογραφηθούν) και άλλοι με εμπειρία σε επιμέλεια κειμένων θα απασχοληθεί [sic] στο Υποέργο 01. (National and Kapodestrian University of Athens Research Board, 2010)

> People *who have English as a mother language* and some kind of theatrical training (in order to play roles in communicative situations that will be audio-recorded) and others with experience in copyediting will be employed in Sub-project 01 [i.e., development of language assessment instruments]. (my translation, emphasis added).

In other words, it would seem that the State Certificate of Language Proficiency also valorised Inner Circle norms in practice, their claims to the contrary notwithstanding.

Another feature common to all the examinations was the communicative definition of the linguistic constructs that they tested. All the examinations primarily measured the candidates' ability to use the language in order to achieve communicative goals. This orientation was most obviously evidenced in the way that all the examinations had been mapped against the Common European Reference Framework, which defines key communicative competencies at various levels of ability (Council of Europe, 2001). In addition, candidates were mainly examined through direct testing of their communicative skills (i.e., through reading, writing, listening, and speaking tasks). The LTE tests, which simulated real-life scenaria, were a prototypical example of such communicative testing, although it should be noted that the face validity (Alderson et al., 1995) of these examinations seemed to suffer from the absence of a grammar and vocabulary component. The UMI/ELI tests, which used discrete-item multiple-choice testing extensively to test grammar awareness, lexical range and the receptive skills, was a notable exception, but it is interesting that these tests were nevertheless described as Examination(s) of *Communicative* Competence/Proficiency in English, thus indicating at least nominal alignment with the communicative orthodoxy.

5.2.3 On hierarchies and determinism

To connect this to the description of the state space which was outlined in Chapter 3 (Figure 3.1), the specifications of the certification examinations seemed to orient certification towards the Standard Language Ideology position in the linguistic dimension, the communicative position in the pedagogical dimension, and – in the absence of any credible evidence to the contrary – the neutral position in the political dimension. The intersection of these three positions overlaps with what I defined as the 'mainstream' constraining structure in the state space (Table 3.1), and it could therefore be argued that the certification intentionality pushed the system firmly towards that direction.

When thinking of the certification intentionality, we might be forgiven for assuming that intentionalities function as mechanisms though which higher-order structures (the effect of international examinations) exert deterministic influences on lower-order ones (such as the small culture of a language school). It is true that the higher-order structures – the mainstream ideologies underpinning ELT – seemed to have had a more dominant role in shaping the orientation of the intentionality, and this appears to be consistent with the processes described in Young et al.'s (2002) ecological model. Moreover, it aligns with concerns raised in the critical ELT literature, such as Phillipson (1992), which adds theoretical validity to this interpretation.

However, a closer look suggests there are interesting deviations from what a deterministic account might predict. For example, even though certification was strongly associated with Communicative Language Teaching, teachers and learners in the language school seemed sceptical about the examinations that had the most strongly communicative format (LTE). This was at least in part due to the fact that the examination did not conform to proximal and intrinsic expectations about the importance of grammatical accuracy, and thus its face validity was challenged. Similarly, a large number of learners indicated a preference for the UMI/ELI exams, which employed a format not commonly associated with communicative language testing. In other words, intrinsic and proximal shaping influences seemed to mediate the effect of more distal ones.

5.3 Finding a place to belong

Another intentionality that drove activity in the language school was *international integration*, a desire to enhance the learners' ability to engage in English-medium communication in a global setting. Deriving from Gardner and Lambert's (1972) distinction between integrative and instrumental motivation, the term 'integration' has been re-connotated here to reflect sociolinguistic changes brought about by the global spread of English. Integration, in sense used in this discussion, does not refer to assimilation in the culture of a host country where English is spoken, but rather to the learners' desire to be accepted as competent members in transnational English-using discourse communities (Widdowson, 2003). Moreover, unlike Gardner and Lambert's usage, the term is not used in juxtaposition to 'instrumental' motivation (the desire to perform tasks through the use of English). Rather, the ability to successfully perform tasks in English is viewed as the criterion that defines membership in English-using discourse communities as 'successful users of English' (Prodromou, 2006, 2008). In this sense, the construct of international integration fuses the two motivational orientations proposed by Gardner and Lambert (1972).

Like certification, the international integration intentionality appeared to emerge from a variety of different intrinsic, proximal and distal influences; but unlike certification, these seemed to lack cohesion and intensity. These shaping influences include a weak 'international-intercultural' orientation of Greek ELT (Sifakis & Fay, 2011), the desire of students at the language school to integrate in various transnational discourse communities, and the discourse norms in these communities.

Looking at the professional community of ELT in Greece, there is some weak evidence of attitudes conducive to the development of the international integration intentionality. In a study that focused on English language teachers in the Greek state school system, Sifakis and Fay (2011) noted the existence of an 'international-intercultural' orientation to ELT, which they define as 'teaching that mostly focuses on preparing learners for English-medium communication in international contexts' (p. 287). Although attitudes towards this orientation were reported to be ambivalent, their findings lend some credence to the claim that that at least some teachers in the Greek context hold views that are compatible with the development of an international integration intentionality.

Sifakis and Fay (2011) stop short of defining what the 'international contexts' might be within which communicative needs are served in English, as that was outside the scope of their research. The data that are presented below, which were derived from a series of qualitative and quantitative surveys in the school, extend their findings, by teasing out three broadly-construed types of discourse communities into which some learners at the language school seemed to want to integrate. Specifically, these are: (a) *vocational-professional* communities, (b) *international youth culture* communities, and (c) a community of *internationally mobile citizens*. I should point out, however, that these discourse communities are conceptualised as overlapping categories, rather than as mutually exclusive groups.

The majority of students seemed keen to obtain membership to what can be defined as vocational-professional communities, for which they perceived English language competence as a requisite qualification. When asked to describe 'why learning English is important for [them]', the student participants typically responded with variants of 'for working in a job which needs English degrees'. In fact, reference to the connection between English language competence and employment prospects was made by almost all the respondents to the surveys, even though most of them were in their mid-teens and were not expected to engage with the realities of the job market in the immediate future. Some respondents explicitly referred to international communication as part of their jobs. Examples of such responses included:

> **Student Quote 5**
> It is useful for most of the jobs nowadays for communication with people from other countries.
>
> **Student Quote 6**
> If in the future you are require [sic] to write a formal letter or talk to somebody older than you, then you will face some problems of communication (unless you attend formal lessons).

Others focused on the societal expectation that future professionals are conversant in English, as was the case in the following response:

Student Quote 7
First of all, learning English is essential enough for our future plans, as jobs are refered [sic]. It is known that nowadays society asks for a variety of qualities and knowing English language is one of them.

It is also interesting to note that the value of English language proficiency was not expressly associated with references to geographical mobility. In other words, the learners did not appear to be communicating a desire to learn English in order to find employment in a country where English was used as an official language. Rather, their responses indexed a belief that the defining feature of membership in the professional and vocational communities into which they aspired to integrate was linguistic proficiency, rather than geographical location.

A second set of discourse communities that seemed to be defined by English language proficiency might be subsumed under the term 'international youth culture'. I use this term to refer to virtual communities of adolescents and young adults who interacted through English or engaged with English-language cultural artefacts. By way of example, many students identified themselves as Beliebers or Directioners (in reference to inexplicably popular singers and bands) or as Twilight enthusiasts and Potterheads (in reference to commercially successful book series and film franchises), and indicated that they often interacted through social media with peers who shared similar interests. For these students, English language proficiency helped them to better appreciate the cultural artefacts around which their communities coalesced, and also facilitated communication among the members of these communities. Some examples of responses typical of a desire to use English for the purposes of engaging with cultural artefacts and discourse community membership included the following:

Student Quote 8
I will be able to see English speaking films without subtitles, because some films cannot be download [sic] with Greek subtitles. Also I can read English books because some books are better when you read them as they were written from English writers.

Student Quote 9
You will chat in a better way, you will be able to understand everything on the Internet and also the songs of the lyrics you listen to.

Student Quote 10
Learning English helps me to communicate with people from another country or understand the lyrics of some songs.

Much like the vocational-professional community, this community also appeared to transcend geographical borders, but it was not necessarily associated with mobility.

The communicative needs of internationally mobile citizens were also mentioned, albeit infrequently, in the context of reasons for learning English. A small number of respondents suggested that they needed to learn English for reasons like the following: 'When I spend my holidays in a foreign country I can communicate with the people over there'. Specific purposes associated with international mobility, such as studying outside Greece were also sporadically mentioned (e.g., 'I also indent [sic] to study abroad so that's another reason, too'). It is interesting that only one respondent explicitly mentioned a country where English is used natively ('travel to the UK'). While being aware of the danger of overanalysing, the lack of such specific references might index some degree of awareness that English is an international lingua franca which need not associated with particular 'native-speaking' countries.

The international integration intentionality could not be readily associated with any pedagogical orientation. Politically, it might be argued that it was oriented towards the uncritical acceptance of Anglophone norms, because many cultural artefacts (music, films, books, etc.) that were important to the international youth culture communities were produced in the Anglophone West. However, what seems important is that the respondents appeared to want to integrate into the transnational community of people who appreciated these artefacts, rather than the specific culture in which these were produced. In terms of its linguistic orientation, it appeared to point towards the ELF position in the state space (Figure 3.1). This was because English was likely conceptualised as a common resource equally available to all the members of the transnational linguistic communities, irrespective of their native and non-native speaker status. Based on the above, it would seem that the international integration intentionality was weakly oriented towards the critical constraining structure in the state space (Table 3.1).

5.4 Teaching about England

A third intentionality, which I termed 'cultural awareness', pertained to expanding the students' knowledge about the dominant cultures of the United Kingdom and, to a lesser extent, the United States (cf. the discussion of cultural flows in Section 3.3.2). To clarify, references to 'UK/US culture' in this section should not be interpreted as naïve to diversity within these communities. Rather, they are used as terminological shorthand to index how salient aspects of these cultures were projected internationally, and how the dominant cultures of the Anglophone West were received locally (at the language school and the Greek society at large).

Although it might be expected that the cultural awareness intentionality stemmed from distal shaping influences, perhaps connected to the Anglophone

West, in fact it appeared to emerge more locally. It seems that the most important shaping influences were Anglophile attitudes, which were particularly noticeable among the language school staff; to a lesser extent, the intentionality also emerged from the (relatively limited) Anglocentric cultural input that was contained in some learning resources. In terms of its orientation, it appeared to align with the technical constraining structure, which prioritised transmissive instruction and a neutral political stance.

5.4.1 Anglophile attitudes

Even though most of the teachers at the language school were of Greek nationality, a recurring theme in many of the teacher interviews was an affinity they expressed towards the dominant cultures of the Anglophone West. One way in which this affinity was manifested was through the expression of positive feelings towards the English language. For example, one of the teachers made it clear that her chief motivation for training as an English Language teacher was that she 'was fascinated by the English pronunciation', which incidentally was why she insisted that her own learners develop 'beautiful' native-like accents. Another teacher was keen to let me know that she was very positively predisposed towards the English language, and volunteered the following information at the end of our interview:

> **Interview Extract 4**
> Achilleas: … we've pretty much covered everything. Is there anything you'd like to add?
> Teacher: About teaching English? Besides the fact that I like the language and I love it and I have loved it for a long time since I started learning it…

A very similar narrative was repeated in yet another interview, where 'love' for the English language was repeatedly cited as a reason for joining the profession ('I suppose I've always wanted to be an English teacher. […] I've always loved the language and I always had a connection to it. Loved it…').

A similar affinity towards the English language was also invoked by several students. In one of the questionnaire surveys, participants were asked why they were learning English, and a number of answers were along the lines of the following response: 'I really like this language and am interested in it'. Elsewhere in the corpus of student responses, the claim was made that English 'is a very interesting language'. These student statements are considerably weaker in strength and relatively fewer than similar sentiments expressed by teachers, but they indicated a positive predisposition towards the language, at least by some students.

The positive attitudes towards English were intertwined with similar affinity towards cultural artefacts of the dominant cultures of the Anglophone West, such as films, songs, and books produced there. The following extract, which is part of my notes from a post-interview debriefing with one of the participants, illustrates the point:

> **Notes extract 1**
> One of the topics we touched upon was [teachers' name]'s love for England. This had not come up quite so strongly in the interview. In fact, there was something in the way she had talked about English Literature earlier on that had given me the impression that Martha was very unlikely to be an Anglophile. However, she spoke very enthusiastically about her visit to London, the friendliness of the people, the architecture… Not only did she want to visit England again, but she'd gladly stay there if she could.

In another interview, a teacher talked extensively about the fact that she had not actually wanted to be a teacher, but she loved Victorian literature, and she felt that by teaching the language, she gave students the ability to access such texts. Another teacher stated that she wanted to attend a postgraduate course in English literature, because in her view 'it is important to know well the culture of the language one teaches'. Yet another teacher claimed that she tried to immerse herself in English culture in order to become a more effective teacher:

> **Interview Extract 5**
> Teacher: I try to I watch movies without, erm?
> Achilleas: Subtitles?
> Teacher: Without subtitles! I listen to English music and try to understand the lyrics. I read magazines, English magazines. I think my biggest problem has to do with the words. I don't know a lot of words. And I try to learn as many as possible. There is a friend of mine who has a book. It's about a game, and has a lot of terms in it, so I borrow it and I try to learn a lot of words.
> Achilleas: What game is that?
> Teacher: It's called D & D, it has to do with Dragons and=
> Achilleas: =Dungeons and Dragons.

The attitudes expressed above were in some ways similar to the data pertaining to the international youth culture intentionality (see above), but differed in two regards. Firstly, the 'target culture' was defined in explicit national terms; and secondly, emphasis seemed to be placed primarily on the valued cultural artefacts, rather than the sense of belonging to a community.

The students' attitudes towards cultural input seemed to confirm a degree of affinity towards Anglophone culture, reductively defined. When I asked a group of upper intermediate learners, in another questionnaire survey, how important it was for them to learn about British culture, 79% of the participants answered that it was 'very important' or 'important' to them. By comparison, feelings appeared to be mixed regarding (US)American culture, which was claimed to be 'very important' or 'important' by two thirds of the respondents, and 'not so important' by the remaining ones. Learning about other Inner Circle cultures, such as Australia or Canada, and about cultures outside the Inner Circle was reported to be even less important. In other words, the learners' responses suggested a hierarchical ranking of cultures, in which British and American cultural input was viewed as more important compared to similar input from other English-speaking communities.

5.4.2 Anglocentric input

Another set of influences that shaped the cultural awareness intentionality related to the learning materials that were in use at the language school. Depending on their cultural content, I classified the activities in the corpus of learning materials in four categories, as follows:

- **Anglocentric** activities were defined as those activities that referred to cultural icons from the countries where English was used as the native language of the majority of the population, such as the UK or the US. Examples included a story about King Arthur and a set of listening activities describing life in Australia. The category also included texts where the country of origin was not explicitly invoked but could be unambiguously inferred, such as a series of passages comparing a cottage in the English countryside, a Tudor house, a boathouse and council housing in a city.
- Topics that drew on contemporary or ancient Greece were categorised as **Hellenocentric**. Examples of such content included a text about the ancient Greek doctor Hippocrates, and a passage about a known Greek actress.
- The **Pluricentric** category encompassed activities that contained cultural input from settings other than the above, such as a series of texts about festivals around the world. It also included activities that seemed designed to raise awareness of cultural diversity, such as a text about English as a Global language.
- Activities which could not be placed in any of the above categories were designated **Neutral**.

In case of ambiguity, priority was assigned in the following order: Pluricentric, Hellenocentric, Anglocentric, Neutral.

The overwhelming majority of the activities that were sampled were Neutral (92.4%). However, when focusing on those activities that were culturally marked, Anglocentric activities outnumbered Hellenocentric or Pluricentric ones by almost 3:1. To be exact, there were 45 Anglocentric activities in the data (4.4%), as compared to 18 Hellenocentric ones (1.8%) and 15 that were Pluricentric (1.5%).

The weak Anglocentric influence in the courseware seems consistent with what is known about the cultural orientation of ELT in the state education system (see Section 1.2 for additional information on context). It has been suggested that, despite claims to the contrary, the official curriculum used in state education appears to aim towards the development of a 'loose sense of cultural awareness about the customs and traditions of the countries where English is spoken as a native language' (Sifakis, Lytra & Fay, 2010). In a content analysis of the textbooks used in state primary education, Pozoukidis and Balabanidou (2010) note that the English language textbooks tend to promote the visibility of 'high-prestige' cultures, a term that they do not unambiguously define, but which can be inferred to be loosely equivalent to 'English-speaking', based on the examples they provide. For instance, they point out that on one of the coursebooks:

> For the overwhelming majority of [named] characters, names are used which belong to high-status cultures (Andrew, Nick, Betty, Carol, Patrick), whereas Greek names (Maria, Sophia, Kostas, Argyro) trail far behind them. But what is particularly important to note is that there are merely four names of characters that come from low-status cultures (Omar, Amir, Yang, Lee), which are mentioned for a total of just ten times in a total of 403 references. (pp. 342–3, translation mine)

Similarly, in Kostoulas (2011) I put forward the claim that much of the content in the ELT textbooks used in state secondary schools promotes a monolithic and somewhat superficial view of British culture to the exclusion of all other communities where English is used.

Taken together, all this information seems to suggest that many learning resources, both at the language school and the Greek context privilege an Anglocentric view of the world. This effect is perhaps weak, since much content is devoid of explicit cultural references, but when cultural imagery is used, it is more likely to refer to the Anglophone West.

5.4.3 The importance of local context

In summary, the Cultural Awareness intentionality indexed the tendency to complement English language instruction with teaching about dominant cultures of the communities with which English was predominantly associated. Importantly, the cultural content with which this intentionality was associated was restricted to British and North American culture, to the exclusion of the cultures present in other communities where English was used. Even within these 'target cultures', the cultural awareness intentionality seemed oblivious to regional, generational and class variations.

The observation that language teaching in the school privileged the dominant culture(s) of the Anglophone West is also consistent with the common trope in the critical literature, namely that ELT is associated with cultural imperialism (e.g., Kumaravadivelu, 2006a; Phillipson, 1992). However, the absence of visible influences directly linking the Anglophone West to the language school adds nuance to our understanding of this phenomenon. Rather than being directly imposed from the Anglophone West, the cultural awareness intentionality seems to have emerged from proximal and intrinsic influences.

Specifically, this intentionality appeared to stem from two sets of influences: Anglophile attitudes, mostly among teaching staff, and Anglocentric input in the learning materials. The role of teachers, in particular, seems consistent with the unverified, but plausible, claim that English language teachers in Greece tend to view themselves as local custodians of the English language and culture (Sifakis, 2009). Although the data in this study can neither confirm nor challenge such a broad claim, it is very clear that the teachers at the language school valued the dominant culture and language of Anglophone countries. Valorising the dominant culture of the UK, i.e., a culture with which the local teachers were highly familiar, might be interpreted as an endeavour to assert their professional status.

The local origins of the cultural awareness intentionality appear to be confirmed by examining the learning materials used at both the language school and state education. It should be noted that the coursebooks used at the language school was produced by local publishing enterprises, or the Greek branches of international publishers, and the fact that textbooks had been 'specially designed for the needs of Greek students' appeared to be a strong marketing point. Similarly, the learning materials used in state education had been commissioned by a government agency, and were locally produced. The existence of Anglocentric content in these materials could be viewed as evidence of pervasive cultural imperialism. However, an alternative argument, which appears to be more consistent with the dialectic relationship between the needs of local teachers and the output of local publishers (Kostoulas, 2007), is that the materials reflected, in part, needs of the local ELT community, including the need to define themselves professionally by virtue of having access to valued cultural information.

In terms of its orientation, the cultural awareness intentionality provides us with a seeming paradox. Although it appeared to derive content from the Anglophone west, its pedagogical orientation was firmly grounded on transmissive teaching practices that were associated with the local educational culture. Specifically, it involved the transmission of a limited set of cultural codes, symbols and values with which local teachers were most familiar. By reducing cultural awareness to information about the 'target culture', the cultural awareness intentionality sustained a power differential between the agents in the system who possessed this knowledge (i.e., teachers) and those who were expected to receive it (i.e., learners); such a power differential is a hallmark of transmissive pedagogy. On the political dimension, the cultural awareness intentionality was oriented towards the neutral position, as there was no evidence of problematisation regarding the effects of cultural flows form the Anglophone west towards the local culture. Going back to the discussion of constraints that were defined in Chapter 3, it seems that this intentionality oriented the system towards the technical constraining structure.

5.5 Being the best

The interaction between the language school and the state education system generated a fourth intentionality, which I have termed competition. This referred to a preoccupation with enhancing learning outcomes for the students who were enrolled in the language school. This was intended as a supplement the state-run ELT provision, which was perceived as being inadequate; however, there was also a concern about making these outcomes visible, in order to demonstrate that the language school provided a valuable service in exchange for the tuition fees that they charged.

The competition intentionality seemed to emerge from a variety of shaping influences, all of which were connected – in one way or another – to the state education system. One set of proximal influences was a set of pervasive societal beliefs regarding the perceived deficiencies of the state education system and the expectations connected to private language institutes. Inside the language school, there was a culture of accountability connected to the performance of students in state education. Finally, the introduction of a Teaching English to Young Learners (TEYL) initiative in the state education system triggered a third set of influences. Although the outcomes of these interactions were not always compatible, a fact that resulted in tensions, the overall orientation of competition appeared to be transmissive.

5.5.1 Limitations of the state education system

The first set of influences that gave rise to the competition intentionality was a widespread societal belief about the inadequacy and ineffectiveness of the provision for teaching English in the state education system. This belief is a recurring theme in descriptions of ELT in Greece (e.g., Angouri et al., 2010; Karavas, 2010; Scholfield & Gitsaki, 1996). Angouri et al. (2010) put forward a tentative explanation for the pervasive lack of trust towards state ELT, claiming that it us due to large, mixed-ability classes, uninteresting lessons, lack of motivation among teachers, and – bizarrely – the fact that tuition fees are not levied by the state schools. Elsewhere in the literature, the ELT provision in state schools is described as inadequate in terms of time, lacking in essential resources and pedagogically inconsistent, and it is suggested that this results in low motivation and what are described as 'behavioural problems' among students (Karavas, 2010, pp. 71–72). Of course, one must acknowledge that all the studies above refer to perceptions of state ELT, rather than actual measurable outcomes. This lacuna in the literature is accentuated by the difficulty in isolating the effect of state education from that of the supplementary tuition (for those students who can afford evening courses) and the effects of low socio-economic status (for those who cannot). Regardless of their validity, however, such perceptions did exert an influence on what students, their parents, teachers and other stakeholders expected out of private language education.

The perception that state education did not cater to the ELT needs of learners was echoed in the responses learners at the language school provided to a set of open-ended questions. When I asked them to compare the lessons at the state schools they attended with those at private language school, the former were generally described as a 'waste of time' and 'irrelevant'. The teachers and the materials were repeatedly described as 'boring', and it was stated that they 'have nothing new' to offer, because the students believed that 'already know most of the things they teach us'. The following response is quite typical of prevailing views about the state education system:

Student Quote 11
All the students are naughty and don't pay attention because they already do private lessons and the teachers can't do their job properly. The books have many errors and mistakes […] we don't have the choice [chance?] to have listening tests, because the state doesn't provide us recordings.

Similar statements were made by almost all questionnaire respondents, including one who added that 'Since they [i.e., the teachers] know that all the students have private English classes, they don't give the appropriate attention to the subject'. In other questionnaire responses, teachers at the state education system were described as 'indifferent', and it was suggested that 'the teacher usually doesn't care if we learn something or not'. In the view of at least one respondent, 'the only thing they [i.e., the teachers] have in mind is how to get done with the lesson, and to finish as quickly as possible the syllabus'. At times, the students' responses appear exaggerated, as was the case with claims that the state school teachers were 'amateur' and that 'some of them do not know English as [well as] some students do'. While occasional overstatement might compromise the trustworthiness of some student-generated descriptions, it seems fairly uncontroversial to assume pervasive dissatisfaction with the ELT provision of the state school system.

Compared to state education, private foreign language institutes, such as the one on which this study focusses, are expected to provide a 'stricter environment with more class tests and greater discipline', better ratios of students / teachers and more exposure to the language (Scholfield & Gitsaki, 1996, p. 126). These expectations were echoed by learners in the language school as well. In the survey mentioned above, learners suggested that lessons in private language centres are 'more difficult', and that students who attend courses in such institutes 'have much studying, but they learn many things'. Elsewhere, it was mentioned that students in private language centres 'are silent, and pay attention to the teacher, because they take lessons seriously'. Lessons, another student added, involve 'very good analysis of [grammatical?] phenomena, multiple examples and exercises'. In other words, students tended to associate teaching and learning in the language school with hard work, high standards, and the kind of classroom atmosphere that is usually associated with transmissive teaching.

5.5.2 A culture of accountability

In addition to proximal influences, such as beliefs about the state school system, the competition intentionality appeared to be shaped by a culture of accountability that had developed among the teachers at the language school. On several occasions, teachers in the school pointed out that they had to accommodate for the emphasis that state ELT placed on grammatical awareness and lexical range, but was unable to foster. For example, in one of the interviews, a teacher complained that she had fallen behind with one of her classes because she had had to make time to help students revise grammar whenever a test was announced in the state ELT classes that they also attended:

Interview Extract 6

Teacher: It's not possible to strictly adhere to the syllabus because, for example, I recently had to dedicate at least half a lesson doing revisions and answering questions for their [state] school.

Achilleas: For their school? Why?

Teacher: ((long sigh)) They… because they were told that they would take a revision test: present simple, continuous, perfect and past, and I had to revise-

Achilleas: Now, I'll take on a provocative role. But shouldn't *that woman* [i.e., the state school teacher] do the revision for *her* test? And you can then go on normally, according to *your* syllabus?

Teacher: The syllabus. Yes. But if someone gets a, I don't know, 12 [out of 20, a mediocre grade], then I will... Me! It will be *me* who will be discussed. They will say that my kids don't even know basic stuff.

It should be noted, by way of providing context, that formal tests in the state education system are regulated by law and that they must contain reading comprehension, morphology and syntax components (see Section 1.2.1). The expectation that students should prepare for such tests by revising grammar was a product of such strong emphasis on form.

Another teacher expressed indignation at what she claimed to be the unfair impression that her students had not developed adequate lexical range to engage with the textbook they were using in the state school. Although she realised that the vocabulary encountered in that textbook was highly technical ('Unless they become aircraft mechanics at age 11, why would they want to know *fuselage* and *ailerons*?'), she nevertheless felt that she was failing her students. She was also concerned that when the students struggled with such language, she appeared incompetent in the eyes of their parents, who may have expected her to prepare the students more effectively. Much like the teacher in the previous interview extract, she complained about the difficulty of striking a workable balance between the requirements of her own syllabus and 'incomprehensible' expectations imposed by the school system. In her view, the only way to cope is to 'give them [i.e., the students] words and explanations, ask them to study and test, test, test'.

In yet another interview, a teacher expressed her surprise that young learners were being made to take grammar tests in English in Year 3 of primary school:

Interview Extract 7
Teacher: The mother of an A Junior [student] told me that the teacher at [the state] school made the girl write tests with present continuous and 'have got'.
Achilleas: That's not part of the [state school] syllabus.
Teacher: That's what she told me anyway, and I was stunned. What does one do then? Explain to her that we are interested in learning how to speak first, right?

This particular extract also shows, perhaps more clearly than the previous two, the tensions between communicative and transmissive ways of teaching. Unlike the teachers in the previous two extracts, this teacher seemed prepared to defend her pedagogical choices ('we are interested in learning how to speak'). However, she appeared to want reassurance from me (whom she perceived as having a certain degree of authority) that she was making the right choices.

What these comments have in common is a preoccupation that if the learners underperformed at their morning classes, this could reflect poorly on the teachers. In response to this, they strove to align their objectives with the strongly accuracy-oriented objectives set by the state school system.

5.5.3 Introduction of TEYL

Another consideration that exercised an influence on the language school was the planned implementation of Teaching English to Young Learners (TEYL) courses in state education. This was due to the fact that private language education and state education in Greece appear to be in a co-adaptive relationship, i.e., a relationship in which changes in one system trigger changes in another one and visa-versa (Larsen-Freeman & Cameron, 2008). In this case, it appears that changes in the state school system were motivated by policies in private education, and then fed back into the latter. Unlike the shaping influences that were mentioned previously, the introduction of the TEYL programme seemed to be orienting the school towards communicative language teaching, due to the difficulty of describing meta-language to younger leaners.

When fieldwork was conducted, English was normally introduced in the state school curriculum in Year 3 of primary schools, but plans were in place for its introduction starting in Year 1, and a pilot programme was implemented in 800 'reformed' primary schools across the country (see section 1.2.1, for additional details). Similar policy decisions had previously taken place in 1995, when English was first introduced in primary education (starting at Year 4) and in 2003, when the starting age for ELT tuition was brought forward to Year 3. The rationale of this particular curricular change was never articulated in a comprehensive and convincing manner, but the official website of the TEYL programme contained an unsigned document, where it was suggested that the programme had been designed to cater to the needs of 'parents who end up sending their children to a language school in order to support their first contact with the foreign language' (University of Athens Research Centre for English Language, n.d., my translation). Additional arguments in support of the TEYL programme were put forward in a series of publications that appeared in online journal set up to showcase the programme (Zouganeli, 2011a, 2011b; Zouganeli & Giannakopoulou, 2011). These vaguely reference the Critical Period Hypothesis (Lenneberg, 1967), the benefits of bilingualism, and examples of other European countries where Primary ELT programmes had been introduced. However, these claims appear to be post-hoc rationalisations rather than arguments that informed the design of the programme.

The Greek TEYL programme, which was piloted between 2010 and 2015, fell short of expectations, and was eventually scaled down. According to a report that was published in *ELT Journal* after the completion of the pilot phase:

> the key stakeholders (parents and head teachers) of this innovation reacted negatively to the introduction of English in the first grades of primary schools. [...] Some believed that their children would confuse their mother tongue with the foreign language, thereby creating problems with the acquisition of both languages. (Karavas, 2014, p. 246)

Moreover, Karavas notes that 90% of the participating teachers had no experience or training in teaching young learners. This resulted in incoherent and often pedagogically unsound teaching practices that undermined the credibility of the programme (Kostoulas, 2015b). Such concerns prompted many parents to turn to private education institutes at an even earlier age, in order to provide more structured language tuition for their children.

The change in school policy was already causing repercussions in the language school. According to the acting Director of Studies, the standard and early enrolment ages to the General English programme had already been reduced by a year, to eight and seven respectively. The changes in the students' cognitive maturity were expected to have a domino effect throughout the curriculum, and necessitated redesigning courses so that they would be more 'children-friendly'. Another effect of this policy was the planned introduction of two Very Young Learners' classes, one of which was targeted at the four- to five-year-old age group, and one for six- to seven-year olds. Learning activities in these programmes made extensive use of teaching methods that aligned with the communicative approach (broadly construed), such as Total Physical Response (Richards & Rodgers, 2014), story-telling (Cameron, 2001) and theme-based projects (Cameron, 2001).

At the time of the study, most of these changes were at the pilot and early implementation changes, and this precluded a definitive answer regarding their potential and limitations. However, the attitudes of language teachers in the school ranged from ambivalent to critical ('You can't have a year's worth of lessons with little songs, clapping hands and ha-ha-ha, no matter what you try'). Questions were raised about the learners' ability to comprehend grammar at such early ages, the extent to which these classes were compatible with the teachers' specialised skills, and the ethics of charging parents for classes 'in which children do nothing but talk to each other' while the teacher looked on. In all, the teachers at the language school seemed to acknowledge that a demanding grammar-based course would challenge the youngest learners, but a strong belief persisted that a traditional approach was preferable to the new communicative courses.

In all, the implementation of TEYL was creating pressure to shift the system towards a more communicative instruction mode, due to the pragmatic realisation that transmissive, grammar-focussed lessons were difficult to implement with younger learners. However, this re-orientation was met with strong resistance, primarily among teachers, who remained unconvinced of the pedagogical value of a communicative syllabus.

5.5.4 Complex interactions, unpredictable outcomes

The Competition intentionality, which emerged from the interaction of the state school system and the language school, referred to the imperative to enhance academic outcomes for learners, so that they will be able to cope with the demands of the state school system. This intentionality emerged from a number of influences, including societal perceptions about the limitations of the state school system, a culture of accountability, and the introduction of the TEYL programme.

In the absence of evidence to the contrary, it can be assumed that the competition intentionality was oriented towards the Standard Language ideology and the neutral political position. The diversity of shaping influences, however, meant that the pedagogical orientation of the competition intentionality was not uniform, but rather it shifted depending on the level of instruction. At the earliest levels of instruction, the repercussions of the planned introduction of TEYL in the state education system oriented the system towards the communicative position of the pedagogical dimension. Later on, however, the accountability culture in the language school created pressure to align with the form-focussed learning objectives that prevailed in the state education system. Similarly, the expectation for rigorous and structured learning activities resulted in the prevalence of transmissive modes of instruction, which allowed for greater teacher control. The tension between these competing effects was manifested in the sceptical way in which many teachers positioned themselves towards the curricular changes that were being introduced.

This complex picture usefully illustrates another property of intentionalities, namely the difficulty of connecting causes and outcomes in linear patterns. In the previous sections, we witnessed how intentionalities tended to have uniform effects on the orientation of the system – certification, for instance, oriented the system towards the mainstream constraining structure of the state space, international integration seemed to be oriented towards the critical one, and cultural awareness was firmly oriented towards the technical constraining structure. The somewhat erratic effect of the competition intentionality, which alternated between mainstream and technical orientations, can only be explained with reference to the mediating effect the state school system, i.e., proximal influences. In other words, although the intentionality remained constant, its effect varied depending on changes that took place *outside* the system. This observation highlights the usefulness of a complexity perspective as an analytical frame that takes contextual influences into account.

5.6 Preventing change

The last of the five intentionalities, protectionism, refers to an implicit agenda of safeguarding the vested professional interests of local ELT practitioners. This intentionality was the outcome of three main shaping influences: (a) perceptions regarding the particularity of the Greek context, (b) the valorisation of the Pedagogical Content Knowledge of locally trained bilingual teachers, and (c) the primacy attached to professional experience as opposed to theoretical knowledge. Although these influences were not always compatible, taken together, they appeared to underscore the comparative strengths associated with the local teachers' professional expertise.

5.6.1 The 'Greek reality'

Key to generating the protectionism intentionality was the discourse, and underlying belief, about what was termed the 'Greek reality'. This discourse indexed the belief that the educational needs of Greek students were different from those described in the mainstream ELT literature.

References to the particularity of the Greek context were ubiquitous. For example, most of the learning materials in use were locally designed in Greece, either by Greek publishers, or by local imprints of international publishing houses (see Section 4.2), and the claim was put forward repeatedly and emphatically that they had been designed to cater to the specific needs of Greek students. For example, the blurb in the back cover of one series read: 'A two-level course for A and B Junior classes, *written especially for Greece,* with lots of exciting resources for students and teachers' (emphasis added). Similar information was highlighted in another set of coursebooks, whose author was described as having extensive experience 'working in the field in Greece'. In the promotional materials of the same series, it was also stated that special treatment was given to 'the common grammatical and lexical mistakes that Greek students often make'. Another series was described by its authors as 'a three-level course for Greek students' and it was claimed that 'it addresses the needs of Greek students and teachers'. Despite the frequency with which the needs of Greek students were invoked, there was little explicit reference in the materials as to what these needs were or how they were catered for.

Some insights about this particularity can be drawn from the interviews with the teachers. In one of those interviews, it was suggested that Greek EFL students tended to differ from their counterparts in other countries because of their age, their linguistic background and the lack of opportunities to engage in authentic communication. The teacher argued that learners in Greece tended to begin ELT courses at very young ages, which meant that they had lower levels of cognitive maturity and different interests, compared to students in different countries. She supported this claim by pointing out that an older (internationally marketed) coursebook that had been used at the language school had proved inappropriate for her classes:

Interview Extract 8
Yes, it was a good book, but it was not very easy to work from with these students we have here. [...] Half the time, I had to explain the ideas in the book. What are the causes of poverty, social problems, volunteering. I had to translate all the time, and they still didn't understand, because [...] they don't know these [concepts].

Regarding the strengths of Greek learners, the teacher mentioned that they benefited from a solid background in grammar, which was extensively and explicitly taught in the state education system. In her view, there were several universal features ('basic grammar phenomena'), which could be transferred from one language to another: she listed the structure of the verb system and the grammatical categories that are encountered in traditional grammar books as an example. As a result, she claimed, it was possible to engage with grammar 'in sufficient depth', which she presumed was not possible in contexts where grammar was only taught inductively. Lastly, she felt that learners in Greece lacked opportunities for authentic communication, which resulted in 'very poor' speaking skills, and hindered the efficient implementation of communicative tasks. For reasons such as the above, she argued that it would be inappropriate to incorporate 'foreign' theory in her lessons.

5.6.2 Locally acquired knowledge

Another influence that contributed to the emergence of Protectionism was the belief that the Pedagogical Content Knowledge of locally trained teachers was particularly relevant to the needs of Greek students. Pedagogical Content Knowledge (PCK) is a broad term that comprises knowledge of context, knowledge of pedagogical theory and subject matter knowledge (Shulman, 1986), but the aspects which were reportedly most relevant to ELT were familiarity with linguistics, explicit knowledge of the grammatical system, and a literary background.

These arguments were often invoked in connection to how such knowledge provided locally trained teachers with an advantage over Native English Speaking Teachers (NESTs). Regardless of their qualifications, local teachers expressed strong convictions that they compared favourably over NESTs, pointing out that the latter often only held entry-level qualifications, and 'become teachers overnight'. When I asked some teachers to explain, in specific terms, the reasons why they felt that they were more effective teachers than NESTs, they usually invoked their knowledge of the formal features of the language and their ability to describe them. For example, one teacher insisted that, having been educated in the Greek school system, which places a premium on metalinguistic awareness in language teaching, she enjoyed significant advantages over any foreigner:

Interview Extract 9
If even I struggle to describe the differences between Present Perfect and Past Simple, or between Past Present Perfect Simple and Continuous, how can an Englishman or an American do this? Let's be serious. […] In England they haven't taught any grammar for the last 20–30 years. In the postgraduate course that I attended, only I and another girl, she was also Greek, knew any proper grammar [and was able] to explain what was happening. The others kept saying things that were not logical, whatever. […] [Native Speakers] are not taught each tense, one after the other. If you ask them 'what is this?' – a Past Perfect for example – they will look at you like fools. They can't even name them [i.e. the tenses].

A similar argument was put forward by another teacher, who was claimed that, despite popular opinion, locally trained teachers were better qualified to teach pronunciation compared to NESTs:

Interview Extract 10

Teacher: And I'm sure you know this better than I do, that – when we were at University we learnt that – for a child to learn, there needs to be some kind of noticing, right? But when an Englishman speaks in correct [pronunciation], in RP [Received Pronunciation], he will say 'dishes'. But because he's a native speaker it will be with devoicing in the d, you know, as in, like tishes.

Achilleas: Do you think that this is a problem?

Teacher: Yes, but- The Englishman does this without understanding. Because he's a native speaker. And not only that, he doesn't even *understand* that this is a problem for the Greek students, if that's how he speaks.

Such sweeping generalisations, and the conviction with which they were expressed, may be interpreted as indications of the local teachers' insecurity, and possibly concerns about the disadvantages they face in a competitive job market, where being a non-native speaker is perceived as a disadvantage. But in addition to that, they also index how an ideology that valorises metalinguistic knowledge can function as an instrument for excluding unwanted competition.

Interestingly, this ideology was weaponised on occasion to reinforce perceived status differences within the language school. Several teachers at the language school had completed four-year undergraduate courses in subjects like pedagogy or the Arts, which – coupled with a C2-language certificate – conferred them a teaching license, but had not trained formally as language teachers. Although the teaching competence of these 'licenced' teachers («επαρκειούχοι») had never been overtly called into question (and it shouldn't!), some teachers who had university degrees in English Language and Literature tended to perceive themselves as being more competent. This attitude was encapsulated in statements such as the following:

Interview Extract 11
These people who are licenced to teach know the language, sure, but we [university qualified English teachers] know how to teach about the language. This is our difference […] Look at how bad things are! Not only are we paid the same salary, but also our knowledge goes unacknowledged. That is, four years [of university studies] mean nothing, and any strumpet who got a proficiency counts for just as much.

In another interview, a different university-qualified English language teacher insisted that 'some of them [i.e., licenced teachers] may know the language well enough', but no teaching license brings them to the level of graduates 'who have devoted their life to studying English'. She also expressed incredulity about how 'four years at university, with literature, linguistics and [ELT] methods' can legitimately be compared to 'a proficiency'.

What these extracts show is, firstly, the prevalence of very strong views regarding the importance of literature and linguistics knowledge, as a requisite of teaching competence. They also show how this ideological belief was brought to bear on discussions that aimed to exclude different categories of teachers, by calling into question their professional adequacy.

5.6.3 Primacy of practice

A third shaping influence was the belief that practical teaching experience constituted a more valuable resource than the kind of received theoretical knowledge that is associated with In-Service Training (INSET) initiatives or advanced studies courses. For example, when I asked about one of the teachers about her professional development aspirations, she argued that she did not feel the need to take an advanced studies course, because '[i]t's not the studies that turn you into a teacher. It's teaching. When I began, I was almost clueless, but I learnt as I moved along, just like my students do.' A similar sentiment, expressed more negatively, was articulated by a different teacher, who appraised an INSET seminar with the following comments:

> **Interview Extract 12**
> The theory was just too much – devastating! – multiple intelligences, affective learning, groups and tasks. I've learnt this already. What I need is advice from someone who has experience teaching classes like mine!

In yet another interview, the following comments were made regarding professional development:

> **Interview Extract 13**
> Achilleas: Moving just for a second outside the agenda of this interview, have you ever considered the courses offered by the [Hellenic] Open University?
> Teacher: That would be a possibility, yes, but I find them a little bit boring to tell you the truth. It's all linguistics. I would like to do something like translation or literature.

Quite interestingly, the belief regarding the primacy of practice seemed to coexist with the ideology regarding the importance of theoretical knowledge that was discussed in the previous section. In fact, in at least two occasions, these two seemingly incompatible beliefs were expressed by the same teachers, who appeared unaware of the self-contradictory nature of their claims. One way to reconcile these assertions would be to suggest that the teachers believed that theoretical knowledge was most relevant in initial teacher education, but not an important component of continuing professional development. Alternatively, it could be suggested that that these contradictions reflected an attempt to define teaching effectiveness in ways that were compatible with the teachers' professional profile at the time of the study.

5.6.4 Incompatible beliefs working together

The protectionism intentionality involved safeguarding the professional interests of local teachers. This was done by valorising the types of knowledge that local teachers possessed, and therefore excluding unwanted competition. The main beliefs associated with the emergence of this intentionality included assumptions regarding the particularity of the Greek context, the ideological belief that a thorough grounding in language and linguistics was requisite to effective teaching, and the belief that professional development should prioritise practical experiential knowledge as opposed to engagement with theory.

The orientation of the emergent intentionality was firmly directed towards the technical constraining structure of the state space (Table 3.1). In terms of the linguistic dimension, it prioritised the standard language ideology, as the teachers – especially those trained in the Departments of English Language and Literature – perceived themselves as custodians of the standard language (Sifakis, 2009). Pedagogically, it pointed confidently towards the transmissive position, due to its emphasis on grammatical form and its reliance on tried-and-tested teaching methods. Politically, it was connected to the neutral position, in that it used the English language and associated technical knowledge in order to preserve an inequitable status quo and covert hierarchies.

From a theoretical standpoint, the protectionism intentionality illustrates how intentionalities can synthesise seemingly incompatible shaping influences. Rather than cancelling each other out, the contradictory beliefs regarding the role of theoretical knowledge appear to work together in order to sustain a set of beliefs about the professional adequacy of the teachers in the language school. Once again, this unexpected finding highlights the analytical utility of a complex systems perspective.

5.7 Dynamics of intentions

This chapter put forward the construct of intentionality, which was defined as a force that drives a system. In the discussion above, I showed how every intentionality is associated with a 'purpose', such as providing learners with certificates, integrating in transnational discourse communities, developing cultural awareness, competing against the ELT provision in state schools, or preserving the structure of the system. To conclude the chapter, I will briefly revisit the properties of intentionalities which were listed in Section 5.1, and connect them to the examples from the language school that illustrate these properties; this will lead to the discussion of one last theoretical construct, the dynamics of intentions.

Intentionalities, I argued, are collective, in the sense that they come into existence from the activity of multiple agents in the system, and they may or may not be present in the mental dispositions of individuals. Each of the intentionalities I described was the product of beliefs and actions of agents inside the system, near it, or at a distance from it. I described these beliefs and actions as shaping influences, and used the terms intrinsic, proximal and distal to denote their relationship to the system. I also noted that proximal shaping influences tended to be more important than distal ones, as we saw in connection to the cultural awareness intentionality (Section 5.4).

We also saw that intentionalities are nested, which means that they can connect higher- and lower-order structures in a system. For example, the desire to integrate in transnational discourse communities seemed to reflect broader social phenomena. However, the relation of the intentionalities to each other and to the system as a whole was not deterministic. In the case of the certification intentionality (Section 5.2), we saw how higher-order structures (the international examinations) oriented the school towards communicative pedagogical practices, but this influence was mediated by the activity of proximal influences.

Another thing we noticed about intentionalities is that they are emergent phenomena. This means that they are created without central organisation. This was most obviously the case with the development of the protectionism intentionality (Section 5.6), which cannot be explained as the product of planning, but had many the hallmarks of organised activity. Because of their emergent nature, they are also sensitive to changes in shaping influences that are part of the system's environment. For example, the competition intentionality seemed to exercise different effects on the language school (Section 5.5). These appeared to connect to changes in proximal shaping influences, namely the policies of the state school system.

Finally, intentionalities are generative, meaning that they create changes in the structure of the system. For instance, the competition intentionality seemed to be driving a reconfiguration of the syllabus in the language school. Also, the certification intentionality appears to have prompted the adoption of communicatively oriented learning materials in some courses (see Chapter 4), and the affordances created by these new materials made communicative language teaching more likely, resulting in another reconfiguration of the system. These generative effects can be associated with different orientations of the intentionalities. Thinking specifically of the pedagogical orientations, we can illustrate these orientations by representing intentionalities as vectors (Figure 5.1). These vectors could be plotted in the state space of the system or the affordance landscape to depict the interplay of intentionalities, affordances and higher-order constraints.

The web of intentionalities in Figure 5.1 depicts the *dynamics of intentions* of the system (Young et al., 2002). This represents the ways in which the intentionalities relate to each other and to the constraining spaces of the state space (the differences in the directions of the vectors have been exaggerated for clarity). As can be seen, the cultural awareness, competition and protectionism intentionalities are oriented towards the technical paradigm, i.e., the constraining space associated with the local educational cultures. In other words, these three intentionalities seem to be pushing the system towards that constraining space, and sustaining the constraining space through their activity. The effect of these intentionalities is flanked by certification and integration. Of these, the former was oriented towards the constraining space that is associated with the global ELT cultures (the mainstream paradigm), but it seems that what drives it is not the effect of professional discourse or teacher training, but rather the influence of examinations. And finally, it appears that the international integration intentionality connects indirectly to the critical paradigm, in part because of the inclusive conceptualisations of the 'target' communities.

Figure 5.1: Dynamics of intentions at the language school

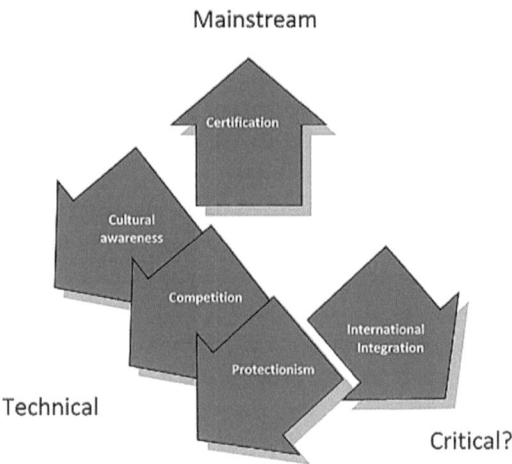

What Figure 5.1 cannot capture is the volatility and dynamism of the dynamics of intentions. This is caused by changes in their relative intensity, and even their orientation, as noted in the description of the competition intentionality. In Section 4.7, I described that the affordance landscape created by the learning materials changed dynamically along three dimensions, which I defined as spatial, phase and diachronic variations. These variations connect to variations in the intentionalities in two ways: on the one hand, affordances in the learning materials can shape intentionalities, and on the other, the selection of learning materials reflects dynamics of intentions. It therefore seems plausible that similar processes of change – from activity to activity, from level to level and from year to year – are also present in the dynamics of intention.

The discussion of intentionalities was an important step towards completing the description of the system. In Chapter 3, I described the state space encompassing all the possible states in the system; this was followed, in Chapter 4 by a discussion of affordances, i.e., action possibilities that made certain states likelier to materialise than others. Adding intentionalities to our conceptual toolkit means that we now have a way of describing what drives the system towards or away from these states. In addition, taking the generative properties of intentionality into account provides us with insights about how the affordance landscape came into existence. In our next step, which will conclude the description of the system, we will look into how the interaction of affordances and intentionalities results in actual activity in the system.

6 The shape of teaching and learning

So far, our developing description of the language school has looked into its state space, the affordances that learning resources created and the intentionalities that emerged within the school. I argued, in Chapter 3, that out of the myriad possible states that the system could potentially take, higher-order structures constrained its activity in three broadly defined regions of the state space, the technical, mainstream and critical constraining structures (Table 3.1). I then showed, in Chapter 4, that the learning resources used at the language school created affordances, and that these affordances increased the likelihood of teaching and learning activities that were associated with transmissive and communicative pedagogy (and, by extension, the technical and mainstream constraining structures). In Chapter 5, I outlined five intentionalities and described the dynamics of intentions that oriented the system towards different states – most commonly, but not always, the technical constraining structure.

In this chapter, I move the discussion forward, by describing the actual teaching and learning activity that emerged from the interaction of affordances and intentionalities. Key to this discussion is the construct of 'attractors', or preferred states in which a system tends to find itself (Thelen & Smith, 1994). For our purposes in this chapter, I will think of the attractor as a sequence of activities that recurred frequently in the language school. Drawing on data from questionnaires, interviews and classroom observations, I will describe three typical attractors, and discuss them in relation to dynamics of intention and affordances that contributed to its development.

I begin this chapter by making a few theoretical comments about the nature of attractors, and by disambiguating some of the terms that I will use in my description of the activity sequences (Section 6.1). The first attractor, which I will describe in Section 6.2, is a sequence of reading and vocabulary activities, which provides insights into the interaction between learning materials, competition and protectionism, and illustrates how pedagogical activity emerged from this interaction. The next attractor, which connected to pedagogically conservative ways of teaching grammar, illustrates how learning materials were reinterpreted, partly due to protectionism, to create a distinctive local form of pedagogy (Section 6.3). Process-based Writing, the third attractor, was located in the mainstream constraining structure, and therefore provides a paradigmatic counterpoint to the other two, highlighting the potential of change inherent in the system and the challenges associated with such change (Section 6.4). The chapter concludes with brief remarks, in Section 6.5, about the processes of stasis and change at the language school.

6.1 About attractors

In complex systems theory, attractors have been defined as regions in the state space of a system 'in which a system tends to move' (Larsen-Freeman & Cameron, 2008, p. 50). For our description of a language school, attractors can be conceptualised as frequently recurring activities, which are associated with specific linguistic, pedagogical and political attributes. For example, if the typical mode of instruction in a school involves presentation, controlled practice and free production of standard language forms, this could be described as an attractor anchored on the technical constraining structure of the state system.

Complexity theory distinguishes between different types of attractors (Larsen-Freeman & Cameron, 2008). When the system remains in the same region of the state space, this is called a fixed-point attractor. The PPP lesson that was described in the previous paragraph is a typical example of a fixed-point attractor. However, sometimes systems can also cycle through a predictable sequence of states, such as when a traffic light periodically goes through a regular sequence of combinations of red, amber and green. This sequence is also an attractor, sometimes called a 'limit cycle' attractor. A hypothetical course that takes place three times a week, during which two communicatively oriented task-based lessons are followed by a transmissive grammar lesson, would be an example of a limit-cycle attractor. Another type of attractor, which will not concern us in this chapter, but is listed here for the sake of comprehensiveness is a 'strange attractor': this involves periods of instability, when the system moves through different states in a seemingly random way.

Learning events, viewed as the contingent outcome of interaction between intentionalities and resources in a class, school or education system can be described as attractors. Although in principle the outcome of these interactions could be unbounded, in practice they seem to be subject to a process of 'social routinisation', which reduces unpredictability and makes them cognitively and psychologically manageable (Prabhu, 1992). In Prabhu's words:

> One needs to be able to anticipate events, in some general form, take some things for granted, even tentatively, and have a frame of reference and roles with which to interpret and respond to what happens. The more recurrent the encounter […] the greater the need for a shared routine and a shared set of expectations. (p. 229)

The existence of such patterns is well documented in the professional literature, where several pedagogical routines are described, including Presentation-Practice-Production (Harmer, 2015), Authentic use-Restricted use-Clarification and focus (Scrivener, 1996), Observe-Hypothesise-Experiment (Lewis, 1993), Pre-task-Task cycle-Language Focus (Willis, 1996), and Before Reading-While

Reading-After Reading (Nuttall, 2005), to name but a few. Although these patterns usually refer to lesson planning, similar sequences are also observable in actual teaching and learning.

In my study of the language school, I used an inductive approach to define seven such sequences (see below), drawing on a combination of observational, interview and questionnaire data and syllabus documents that were made available to me. More specifically, I observed 16 lessons, which I then reconstructed from observation notes and used as a basis for subsequent interviews. Other lessons were reconstructed from lesson plans, lesson outlines in questionnaires administered to the students, and examples of lessons that were described by teachers in interviews. I then clustered these lessons on the basis of similarity, and described prototypical instructional sequences that contained most of the features present in the sequences in each cluster (Geeraerts, 2006; Rosch, 1999). In other words, the descriptions that follow do not refer to specific lessons that actually took place, but rather are idealised abstractions akin to schemata (Lakoff, 1999).

6.1.1 Terminological ground-clearing

In the interest of terminological clarity, before embarking on the description of lesson prototypes, in this section I define a number of potentially confusing terms that are used in this chapter:

- An attractor, or prototypical sequence, is of a series of pedagogical activities that were structured around a specific text or objective. For instance, a Reading and Vocabulary sequence would normally be structured around a textbook passage. Similarly, a Traditional Grammar sequence might be structured around a specific language structure (e.g., 'Learning the Past Simple: Affirmative'. It connects conceptually to a set of constraining structures, affordances and intentionalities, and emerges from their interaction. It may overlap with one of the constraining structures defined in Chapter 3 (in which case it is fixed-point attractor) or it may refer to a periodic movement between more than one such structures (in which case it is a limit-cycle attractor). In the discussion that follows, I will use the term 'attractor' when I want to draw attention to the connection to the state space, and 'prototypical sequence' when I want to refer to the actuality of teaching and learning, but the two terms refer to the same construct. Prototypical sequences, or attractors, are identified in the text through capitalisation (e.g., Reading and Vocabulary).
- Prototypical sequences are divided into several stages. Each stage involves a specific, distinct pattern of interaction among the teacher, the learners and the learning materials. For instance, a Revision instructional sequence might

be divided into stages when the teacher elicited information (*review*), stages when students engaged with exercises (*practice*), a testing stage (*test*) and a stage when feedback was provided (*feedback*). In this chapter, when a word such as *practice* denotes a stage, it is italicised.
- Each stage might contain one or several activities. To illustrate, during the *practice* stage of a Review sequence, learners might sequentially engage with more than one grammar exercises. The term 'activity' is used for terminological consistency throughout the chapter as an umbrella term that encompasses various types of exercises and tasks. In this sense, the term describes modes of behaviour, rather than textual units as was the case in Chapter 4, but there is a one-to-one correspondence between the constructs that the two uses of the term denote.
- The term lesson is used in this chapter to refer to a single class session (usually lasting 50 or 90 minutes), and is therefore distinct from the lesson-as-textual-unit described in Chapter 4. Sometimes instructional sequences would take up a single lesson, but this correspondence was imperfect. It was quite common, in other words, for sequences to extend beyond a single lesson, with some stages being assigned as homework, or being continued into a subsequent lesson. Less frequently, a lesson might contain more than one instructional sequences.

6.1.2 The shape of teaching and learning

As I mentioned above, I reconstructed seven attractors in the teaching and learning activity at the school. These are listed below:

a. Reading and Vocabulary sequences, which were structured around textbook passages, mainly aimed at developing reading comprehension skills and enriching the learners' lexicon. This attractor, which connected to the technical constraining structure, was active throughout the curriculum, being present at Junior, Senior, Upper Intermediate and Proficiency programmes. The Reading and Vocabulary attractor is presented in more detail in Section 6.2.

b. Traditional Grammar sequences focused on learning about one or more language structures, and applying this knowledge onto language use. They were especially common in the earlier stages of instruction (the Junior and Senior programmes), although they were sometimes encountered in the Upper Intermediate programme as well. The Traditional Grammar attractor was typified by a strongly transmissive orientation. A detailed description of a Traditional Grammar sequence can be found in Section 6.3.

c. Inductive Grammar sequences, which were common in the Upper Intermediate and Proficiency programmes, focused on specific language structures, but usually aimed at revising previously-taught forms. In addition, these sequences provided opportunities to develop examination-related competencies by applying metalinguistic knowledge on examination style activities. This attractor differed from Traditional Grammar (above), in that it seemed to be informed by cognitive psychology and constructivism, and it was less dependent on teacher-generated input. Although it was not unambiguously anchored on the mainstream constraining structure, this attractor was closer to communicative teaching than was the Traditional Grammar attractor was.

d. <u>Genre-based Writing</u> sequences aimed at familiarising learners with the formal features of various text genres, such as stories or letters of complaint. The Genre-based Writing attractor was anchored on the technical constraining structure, due to its transmissive pedagogical orientation: students were provided with appropriate information and scaffolding, which aimed at fostering their ability to produce specimens of the genre on their own. Genre-based Writing was relatively common in the older programmes, i.e., Junior and Senior.

e. <u>Process-based Writing</u> sequences also aimed at fostering the learners' ability to produce written discourse, but the learning objectives with which they were associated involved developing specific writing sub-skills (e.g., planning, editing). The Process-based Writing attractor was firmly anchored on the mainstream constraining structure, based on its communicative pedagogical orientation. It was mainly encountered in the Upper Intermediate and Proficiency programmes, and additional traces were in evidence in the Senior (New) programme. The attractor is described in more detail in Section 6.3.

f. <u>Integrated Oral Skills</u> sequences were thematically linked collections of activities that aimed to develop the listening and speaking skills. They typically started with transmissive listening activities, followed by communicative speaking tasks. Both types of activities (listening exercises and speaking tasks) emulated common examination tasks. In other words, they were an atypical example of a limit-cycle attractor that alternated between the technical and mainstream constraining structure. Integrated Oral Skills sequences were fairly uncommon in the early stages of instruction (the Junior and Senior programmes), but their frequency increased in the Upper Intermediate and Proficiency programmes.

g. <u>Reviews</u> were multi-lesson sequences, in which previously learned grammatical structures and vocabulary would be revised, a test would be administered and feedback would be provided to the learners. This attractor, which was unambiguously anchored on the technical constraining structure, was ubiquitous across the curriculum, although the format and formality of the tests varied from level to level.

Table 6.1: Prevalence of attractors in the curriculum

	Junior	Senior (New)	Senior	Upper Intermediate	Proficiency
Grammar	Traditional Grammar			Inductive Grammar	
	Review				
Vocabulary	Reading & Vocabulary				
Reading					
Listening				Integrated Oral Skills	
Speaking					
Writing		Process-based		Process-based Writing	
	Genre-based Writing				

The prevalence of attractors in the curriculum of the language school is presented in Table 6.1. Rows represent various strands of the curriculum, and columns represent the main programmes of study, ranked from the earliest (Junior) to the most advanced (Proficiency). Attractors have been colour-coded: dark grey represents attractors that were associated with the technical constraining structure, and light grey is used to index an association with the mainstream constraining structure. In interpreting this table, it should be noted that the absence of information in certain cells does not indicate neglect or imbalance in the curriculum, but rather that the 'missing' skills were integrated in other sequences, or were taught independently as minor stand-alone sequences, unrelated to the seven attractors that are described in this chapter. In other words, the prototypical sequences described in this chapter do not represent a comprehensive description of all the instructional activities that took place at the language school, although they provide very extensive coverage.

After this overview, I will now go on to present a more detailed description of three attractors, which have been selected based on salience and theoretical significance. The Reading and Vocabulary attractor, which will be presented first, is especially important because of its prevalence. As can be seen in Table 6.1, this sequence was encountered across all levels of instruction at the language school. Traditional Grammar and Process-based Writing sequences were restricted to the upper and lower end of the curriculum respectively, and their complementary distribution helps to illustrate the dynamics at different phases of the school's curriculum (see Section 4.7).

6.2 Reading and Vocabulary

The Reading and Vocabulary instructional sequence was succinctly, if somewhat incompletely, described by a student as follows:

> We read the story of the text, and the teacher asks us questions and then [we do] the vocabulary activities. Sometimes she makes us translate the text to show that we know what the words and the text means, and if I don't know a word I stop and ask her to explain it for me.

This attractor was the most salient one in the language school in the language school. In fact, its prevalence was such that in the discourse of the language school, these sequences were often referred to as 'the lesson' («μάθημα»), suggesting that they were perceived as a kind of 'default' sequence. In Reading and Vocabulary sequences learners typically engaged with a passage that was contained in their coursebooks (sometimes called 'the story'), working under the direction of the teacher. These sequences appeared to serve multiple goals: written passages were often read for pronunciation practice, information was extracted from the texts to develop reading comprehension, and useful vocabulary was highlighted to enrich the learners' lexicon.

Table 6.2: Overview of a Reading and Vocabulary sequence

Stage	Description & comments	Timing	
Prompt	• Teacher pre-teaches vocabulary, activate schemata, and/or establish continuity	Timing for these stages varied, depending on the students' level and the lexical density of the passage	Lesson 1
Read & Listen	• Students read a text while listening to a recording. • Sometimes, simple comprehension task		
Read aloud	• Students read passage for pronunciation practice • Incidental vocabulary explained		
Reading comprehension	• Comprehension questions answered		
Vocabulary consolidation	• Detailed explanation using handout • Copied to vocabulary notebook (homework?)		
Vocabulary practice	• Several multiple-choice, matching, etc. activities • Often based on workbook		Lesson1 Homework?
Dictation	• Simple test focusing on spelling and recall.	10–15'	Lesson 2

Prototypically, a Reading and Vocabulary sequence involved seven stages, spread out over two lessons (Table 6.2). After an introductory 'prompt', learners would silently read a passage while simultaneously listening to it. Then, they took turns reading the text aloud, at which point unknown lexis was explained by the teacher. Next, the students engaged with such 'reading comprehension' activities as might be included in the textbook, and following that they 'consolidated' vocabulary knowledge. Additional vocabulary work, which was sometimes assigned as homework, involved copying vocabulary in dedicated notebooks, and engaging in 'practice' activities. The Reading and Vocabulary sequence concluded in the next lesson with a 'dictation', where the retention of vocabulary items was assessed.

6.2.1 Prompt

As stated above, the Reading and Vocabulary sequences often began with some kind of *prompt*, which was sometimes labelled 'pre-reading' in the lesson plans. Prompting stages, where present, seemed to serve one or more of three functions, namely pre-teaching vocabulary, activating schematic knowledge, and establishing narrative continuity.

Sometimes teachers used the prompting stage in order to pre-teach lexical items that were considered critical for text comprehension. The following lesson observation notes how a teacher used deixis and realia on order to elicit the meaning of key vocabulary items from her A Junior students:

> **Notes Extract 2**
> Next, the teacher took out a set of flashcards and showed them to the learners, asking them what they show. The first was a hamster. A student pointed out that «είναι χάμστερ, αλλά δεν ξέρω πώς το λένε στα Αγγλικά» ['It's a hamster, but I don't know how it's called in English']. Another added that she had a hamster at home, so the teacher took the opportunity to ask what colour it was:
> S: Καφέ και άσπρο ['brown and white'].
> T: OK, στα Αγγλικά ['in English'] brown and white ((points to the next flashcard, which shows a brown blot, and the white walls)) BROW::N . and . WHI::TE. Is your hamster SMA::LL? [gestures]
> She used her flashcards and similar phrases to elicit the meaning of several words, and then proceeded to show the flashcards around very fast for the students to identify.

Some other pre-teaching activities that were occasionally mentioned in the lesson plans included the generation of mind-maps or the completion of cutout diagrams. These were commonly carried out by the students working individually, and then feedback was provided in plenary mode. Pair- and group-work were occasionally mentioned, but these modes of work were not common, as they were perceived to be inefficient, and because students tended to use Greek when collaborating.

A second function of the prompting stage was to activate cognitive schemata, or to equip the learners with the background knowledge necessary for engaging with the passage. In one Upper Intermediate class, for instance, prior to reading a text about Bonfire Night, the teacher elicited information regarding a similar local custom and went on to deliver a brief history lecture on the Gunpowder Plot. In other lessons, students were called on to predict the content of the passage, as seen in the following extract:

Notes Extract 3
The teacher asked the learners to see the pictures [i.e., the illustrations of the cartoon] and guess the plot. She used phrases like «Τι συμβαίνει;» ['What's going on?'] and 'What's this?', 'Who's this?'

A more elaborate prediction activity, from a different lesson, is described below:

Notes Extract 4
The teacher explains that they will read a text about a yacht the crew of which mysteriously disappeared. The class is divided into four-member groups […], and they brainstorm possible explanations […] Representatives from each group are called to the board to write their ideas […].

The extract above is of interest in that deviated from the teacher-centred norms that seemed to typify prompting, and indeed the entire instructional sequence.

Some coursebooks were structured around continuing storylines, and in these cases, the prompting stage was often used by teachers to make narrative connections between previous lessons and the passage with which the students would subsequently engage, as seen below:

Notes Extract 5
[…] the teacher asks a student to summarise the Anna Frank text [which had been read in the previous lesson]. [The student] achieves that with minimal prompting, which mostly serves to keep him from going into unnecessary detail.

The use of 'display questions' (Long & Sato, 1983) strongly resembled oral examination techniques typical of traditional pedagogy, which relied on the rote reproduction of pre-specified content, and in fact they were occasionally described by students in similar terms (e.g., «λέμε το μάθημα» ['we narrate the lesson']). From the teachers' perspective, these activities were said to facilitate transition, and to provide communicative practice.

In general, prompting stages tended to be rather brief (rarely exceeding ten minutes), and as can be deduced from the examples above, they were usually teacher-directed in the interest of efficiency.

6.2.2 Reading and Listening

The next stage of a Reading and Vocabulary instructional sequence was often referred to as '*read and listen*', after a commonly encountered textbook rubric. During this stage, the students engaged with the text by reading silently while listening to a recording of the passage. Sometimes, the text was read out by the teacher for the benefit of the learners, either because a recording was not available, or because the native speaker pronunciation in the recordings was considered too challenging for the learners. Such a sequence is described below:

> **Notes Extract 6**
> Then [the teacher (name redacted)] played the recording that accompanies the story and instructed the learners to 'read and listen'. I was struck by the fact that many learners (though not all) were tracing the sentences with their fingers as they were being read […] The recording was repeated once.

The rationale behind this procedure, according to a senior teacher, was to expose learners to the 'correct pronunciation of words' so that they will be able to reproduce them accurately later. This belief was grounded, she continued, on the principle that the receptive skills must precede production.

Although the students were often expected to passively listen to the recording while reading, on some occasions they were also tasked with answering simple comprehension questions. The rationale behind these unchallenging questions did not appear to be reading comprehension as such: rather they appeared to serve the function of sustaining the students' motivation in the face of an activity that often failed to stimulate much interest otherwise. This can also be deduced by the fact that, more often than not, answers to these comprehension questions tended to be found towards the end of the passage, and were not always key to its comprehension.

6.2.3 Reading aloud

Following *read and listen*, learners took turns to read the text out loud. This stage was generally referred to as 'reading', but will use the term *reading aloud* to designate it, because of the potential ambiguity of the former term. It appeared that one of the main functions of this stage, especially among the less advanced learners, was to practice pronunciation. This goal is visible in the following extracts from my classroom observation notes:

> **Notes Extract 7**
> Following that, [the teacher] played the recording again, pausing at the end of each sentence and having it repeated by one of the learners. This was done three times until all the learners had read several different sentences. Finally, she assigned a different role to each learner and had them to act the cartoon out (while remaining at their seats, though).

> **Notes Extract 8**
> Then the students open the books and take turns reading the text, with the teacher occasionally correcting various words they mispronounce – and there are a few of them. These corrections are generally ignored by the students, who read on unperturbed, which makes me wonder whether they are for my benefit.

It is interesting to note that, even though students did not always take much heed of feedback, the teacher in the second extract seemed conscious of an expectation to correct non-standard pronunciation during this stage of the lesson, hinting at the prevailing valuation of accuracy.

The *reading aloud* stage also served as an opportunity to explain such unknown lexis as might be present in the text. This function is illustrated in the following extract, which describes part of an Upper Intermediate lesson:

> **Notes Extract 9**
> Following that, the teacher nominated students to take turns reading different paragraphs in the text. At the end of each paragraph, the teacher asked the learners if they had unknown words, and when they did, she first elicited responses from other students and, failing that, provided a definition in English. Some learners requested confirmation by providing the Greek equivalent, and the teacher either nodded or provided an alternative definition. Most learners seemed to gloss the words on their books, although some were apparently taking notes on a vocabulary notebook.

This process was similarly described by students, according to whom 'when [the teacher] founds in the text an unknown words she explains us the meaning' [sic]. The preferred language for conducting this activity was said to be English (e.g., 'My teacher explains the words in English but rarely explains them in Greek.'), but this was not always confirmed during the classroom observations. The discrepancy between statements and observed practices hints at both the strength of the monolingual policy of the language school and the challenges of constantly implementing it when engaging with tasks that made multiple demands on the learners' cognitive resources.

This stage was described in different ways by teachers and learners. The students gave fairly straightforward descriptions of how they engaged with the vocabulary, such as the following: 'The teacher tells me which words I don't know, and if we don't understand and in Greek also', or 'The teacher explains the words that I do not know, and if we don't understand the explanation, she repeats the explanation in Greek'. Teachers, on the other hand, tended to describe somewhat more elaborate processes, and allegedly encouraged learners to be active and use English as much as possible. For instance, one of the teachers claimed that she relied on elicitation techniques to capitalise on the pre-existing knowledge of her students:

Notes Extract 10
Since they hear many words in their primary school as well, I give them the word and I ask them if they understand: 'this- who knows this word?'. Or if they don't know it, I will explain it in English.

Similarly, another teacher claimed that she encouraged her learners to infer the meaning of various lexical items:

Notes Extract 11
If they [i.e., the students] do not know a word, then I will say it, or sometimes I will have a classmate who either knows the word from somewhere, or guesses what it means from the co-text. Because they must learn, and it is important [to learn] how to guess what unknown words mean. That is, if you don't have a teacher on standby to ask all the time.

More often than not, newly taught lexis was recorded directly on the coursebook using its semantic equivalent in Modern Greek. Although this practice was ostensibly not in line with the monolingual policy of the language school, it was in fact observed in a number of lessons, as seen in the following notes:

Notes Extract 12
Most learners seemed to gloss the words on their books, although some were apparently taking notes on a vocabulary notebook.

Notes Extract 13
The words were then identified and glossed in the handout (where they were listed alphabetically by unit).

Further confirming its prevalence, the use of glossing was also attested in the students' questionnaires. In a questionnaire survey, I asked a group of Upper Intermediate students how often they wrote down the meanings of unknown words on their book. They all responded that they did, with the majority (55%) answering that they did so very often.

Once again, there seemed to be a discrepancy between the observational data and the information provided by the students, on the one hand, and the teachers' descriptions on the other. In one of the interviews, a teacher insisted that glossing was only sporadically used as an 'emergency' technique for dealing with vocabulary of secondary importance, which was not otherwise available to the students:

> **Notes Extract 14**
> In reality, I find the vocabulary list that we give them [i.e., the students] in every chapter to be very handy. That is, we can go there and say the meanings of the words, and the children can find them there. And if there are other unknown words, which are not in the list, we can make a note in the book, so that they know what it means, but it's not part of the syllabus.

Once again, the overall impression that I formed was that the multiple emphasis (reading, pronunciation and vocabulary learning) in this stage appeared to overburden the students' cognitive abilities. This resulted in the use of relatively straightforward learning strategies, which were tacitly accepted by the teachers. However, such strategies seemed to be frowned upon as they deviated from the monolingual policy of the school, and it is telling that teachers tended to downplay their prevalence.

6.2.4 Reading comprehension

In the next stage of a prototypical Reading and Vocabulary instructional sequence, *reading comprehension*, students focused on the propositional content of the passage around which the sequence was structured. This typically involved skimming and scanning to engage with the reading comprehension activities that typically followed the passage. As before, there seemed to be a strong proclivity for conducting these activities under tight teacher control. In almost all cases, the reading comprehension activities were completed by the students working individually, possibly under a time limit. This process is illustrated in the following extract from classroom observation notes:

> **Notes Extract 15**
> The students take turns reading the multiple-choice questions after the text, and the teacher instructs them to look back into the text for the answers. She insists that they should pay special attention to words like 'no, never, rarely, seldom', because they are 'misleading'. They should also underline the relevant extract in the text.

Following the completion of the task, the teacher would normally nominate students to read out the correct answers. Alternatively, answers might be displayed using an overhead transparency or the interactive whiteboard. The reading comprehension tasks were sometimes assigned as homework, but this deviation from the prototypical lesson was relatively infrequent.

6.2.5 Vocabulary consolidation

Once the passage had been processed for meaning, the next stage of the Reading and Vocabulary instructional sequence involved revisiting its vocabulary content. Unknown lexis was sometimes pre-taught (see Section 6.2.1) or, more usually, explained when first encountered during the *reading aloud* stage (see Section 6.2.3), so this stage of the lesson involved more focused work aiming at the *consolidation* of newly acquired lexis. To that end, students might be provided with wordlists containing the vocabulary items to be learnt (see Chapter 4) and / or asked to record the lexis in dedicated vocabulary notebooks.

The use of these wordlists was observed in several lessons, and was attested by many students, as seen in the following examples:

> **Student Quote 12**
> In each unit my teacher give [sic] me a list of words that are used in this unit.
>
> **Student Quote 13**
> I have to learn about fifteen (15) words in each lesson. They are chosen by my teacher and then she gives us a list with these words and their English meaning.
>
> **Student Quote 14**
> I have 12 words to learn in each lesson. These words are chosen by our teacher.

Normally, these lists would be perused in class, possibly read aloud, and the teacher would provide clarification as required. In some of the classes that I observed, students persistently asked for the Greek semantic equivalent of each item in the list, which they then proceeded to record next to the definition. Interestingly, teachers appeared to be reluctant to engage in direct one-to-one translation. In the words of one learner: 'if we don't understand the meaning of word she explain [sic] us again and again', but sometimes the equivalent would be helpfully suggested by one of the other students. Some teachers suggested that they would on occasion resort to using Greek but 'only if I understand that they [i.e., the learners] are way off' in their comprehension of the English definition. Greek also tended to be used on those occasions when significant time constraints necessitated that this stage be conducted rapidly.

Table 6.3: Format of vocabulary notebooks

#	Programme	Vocabulary Notebooks
II.4	Junior	[The teacher] then […] told the learners to write down the new words in their vocabulary notebooks under headings such as 'nouns' and 'adjectives'.
II.1	Senior	The students were asked to open their notebooks and so that [the teacher] could inspect whether they had copied the vocabulary from the previous lesson.
II.3	Upper intermediate	The words were then identified and glossed in the handout (where they were listed alphabetically by unit), and the students were instructed to copy them in a separate (thematic) section in their vocabulary notebooks titled 'Jobs and professions'.
II.2	Upper intermediate	Some were apparently taking notes on a vocabulary notebook, presumably using personal recording systems

A vocabulary notebook, onto which new lexis was recorded, was also used in most classes. Table 6.3, reconstructed from my classroom observation notes, shows how the vocabulary notebooks were used in various classes, and illustrates the variety of possible formats.

The importance attached to the notebook as a learning aid can be attested by the fact that students were normally required to record all new lexical items, even when these were available in other equally accessible resources, such as the wordlists referred to above. Copying the content of the wordlist onto the vocabulary notebook was usually assigned as homework. For one of the teachers, assigning such copying tasks ensured the students' engagement and facilitated the retention of lexis ('if they just retain half of it, it's still good for them'). More pragmatically, another teacher stated that, in her experience, students often lost handouts, so it was useful for them to have a well-bound, lasting master wordlist at their disposal.

Perhaps unsurprisingly, the students' attitudes towards this practice were ambivalent. Some claimed that the process of creating such a notebook served an important mnemonic function. According to some questionnaire responses:

Student Quote 15
Our teacher give [sic] us a photocopy of some words and she asks from us to copy them in our notebook. I think it a nice way learning new words because as you copy you learn them well.

Student Quote 16
I find it OK, because we learn the words well and as we copy we just learn better and we can easy remember them.

However, the point was also made that mechanical reproduction was a counterproductive way to learn:

Student Quote 17
My teacher give [sic] us a photocopy to copy in our notebooks so we'll learn them. I don't find it helpful, ~~because~~ sorry about that but I find it silly to copy them, it's a waste of time. We'll learn them better only from the photocopy.

Student Quote 18
Sometimes when you copy you don't concentrate on the words and it's a waste of time. I think that the teacher should suggest us some ways we can learn new words because as you know, people have different ways to learn them.

Some vocabulary notebooks were highly personalised, as teachers encouraged students to illustrate their notebooks according to their tastes and interests. Nevertheless, both the content of the notebooks and the process of populating them with words were indicative of the strong control teachers exercised on the learning process.

6.2.6 Practice

To further enhance vocabulary retention, the new lexical items were typically used in exercises, such multiple-choice, gap-and-cue and other similar items. In these exercises, emphasis was placed on the accurate reproduction of orthographic form, and on awareness of each item's semantic content. In most of the lessons I observed, students tended to engage with these exercises individually during class, or they were tasked with completing them at home. After the exercises had been completed, students would usually take turns reading out their answers, which the teacher would either acknowledge or correct. Alternatively, the correct answers might be provided on the whiteboard or using an overhead transparency. In general, these activities seemed to be done rapidly, probably due to time constraints.

6.2.7 Dictation

The final stage of a Reading and Vocabulary sequence was a *dictation*, which took place in a second class session. It should be noted that term 'dictation' was used in the discourse of the language school somewhat loosely to include any type of assessment activity that was conducted individually by the learners and subsequently marked. This could include traditional forms of dictation, as described in the following interview extract:

Interview Extract 14

Teacher: The students are mainly tested through a dictation, […] I do a dictation, a summary of the text we did that includes these words.
Achilleas: And they must…?
Teacher: They must write them correctly. If time permits, two or three sentences will be translated into Greek. So that we know what's going on.

Alternatively, 'dictation' could refer to any of kind of vocabulary-focused task that was marked, as seen in the extracts below:

Interview Extract 15

In the next lesson they write a dictation, which I create on my own, and I include those words that I consider to be the most important. The exercises I use are matching synonyms or antonyms, writing sentences with some words I give them, completing gaps and replacing phrasal verbs with verbs that have the same meaning.

Notes Extract 16

The students are instructed to turn [the sheet containing their homework] over so that they can write a vocabulary test on the back side of the sheet. The test consists of five words which are dictated and have to be defined in English, five definitions for which the students need to provide the word, and five verbs for which they are required to produce a derivative noun.

Differing attitudes were documented regarding dictations: in their interviews, teachers tended to acknowledge that these tests served a direct pedagogical function as students 'internalised' vocabulary through practice, and an indirect one in that they 'encouraged systematic studying'. Some also spoke of the need for accountability, which was catered for by dictations. On the other hand, a number of complaints were voiced, including the demands that these tests made on preparation and class time. Doubts were raised by one teacher regarding the long-term benefit of memorisation ('Both I and the children hate dictations. Vocabulary is learnt through familiarisation and use rather than by rote learning'). A similar sentiment was echoed by one of the students: 'An another [sic] thing that I don't like very much was the small vocabulary tests that we write every day'.

6.2.8 Emergence in teaching and learning

As suggested by its name, the Reading and Vocabulary instructional sequence broadly consisted of a series of stages involving the pedagogical exploitation of a passage, followed by another series of stages focusing specifically on its lexical content. The transmissive practices that were documented throughout the sequence place this attractor firmly in the technical constraining structure of the state space defined in Chapter 3 (Table 3.1). This is an important observation, because by distancing itself from the mainstream structures of global ELT, the language school seemed to operate in a niche where local expertise was at a premium.

In addition to its descriptive value, the discussion above also provides us with insights into the ways in which this attractor emerged from interaction between intentionalities (particularly protectionism and competition) and affordances in learning materials. In Chapter 4, I noted that the materials of different syllabus strands seemed to be associated with different pedagogical orientations. Vocabulary and reading activities usually had a very strong transmissive orientation, especially in the Senior programmes (see Figures 4.2 & 4.3). In light of this, the transmissive modes of teaching that were typical of this instructional sequence can be viewed as evidence of the ways in which practice is influenced by affordances in the learning materials. Additional evidence of this phenomenon can be found in the strong structural resemblance that the attractor bears to the organisational structure of coursebooks in the Junior and Senior programmes (Tables 4.1, 4.2 and 4.3).

One might hypothesise that the attractor was produced by mindless implementation of coursebook content, but a closer investigation suggests that a more elaborate process was in place. In the Upper Intermediate programme, the organisational structure of the learning materials was quite different (Table 4.4), and the affordance landscape also changed to indicate a higher prevalence of learning activities with a communicative pedagogical orientation. Nevertheless, the Reading and Vocabulary attractor was present in these levels as well, and there were no notable deviations in its manifestation. This suggests both the resilience of the attractor, which remained in place even when materials were used that were not optimally suited to its implementation. It also hints at the existence of other forces that sustained the attractor, in addition to any affordances that might or might not be present in the learning materials.

This observation brings us back to the discussion of intentionalities, of which protectionism seems particularly relevant to the creation of this attractor. Traces of protectionism can be found in the ways in which the Reading and Vocabulary instructional sequence deviated from mainstream pedagogical orthodoxy. In the professional ELT literature, reading lessons are commonly divided into pre-reading, reading and post-reading stages. The Reading and Vocabulary instructional sequence conformed to this pattern very broadly, but differed in that the post-reading stage was substituted by extensive vocabulary-focused work. In doing so, teaching and learning emphasis shifted from language use to language form, which is a hallmark of the protectionism intentionality. Other traces of protectionism were to be found in the transmissive ways in which reading passages were put to pedagogical use, and especially during the *reading aloud* stage. The deconstruction of the text in small segments for pronunciation practice seemed related to the valuation of accuracy, which typified local pedagogy. Similarly, the ways in which the segments were processed for meaning, often by means of translation, was indicative of how shared knowledge of the local language impacted practice.

In addition, the Reading and Vocabulary attractor appears to connect to the competition intentionality. A very visible trace of competition was in the *dictation* stage, which invariably concluded Reading and Vocabulary sequences. This stage seemed to be underpinned by considerations of accountability and academic rigour, which were highly valued in the language school. Similar considerations appeared to be motivated by the language school's rather inflexible monolingual policy. As it was felt important that all activities must be carried out in English, monolingual wordlists were normally provided for the students' reference, and the use of translation as a teaching technique was expressly discouraged. Although observational data suggested that monolingual instruction put strains on the learners' linguistic and cognitive resources, and thus resulted in inefficiency, it was interesting to note that deviations from this policy were downplayed, or not acknowledged at all, especially by the teachers, who appeared to be more sensitive to the implications these deviations might have on their professional standing.

In all, the Reading and Vocabulary attractor was one of the most salient features of instruction at the language school. This was associated with extended sequences, which served multiple teaching goals: alongside the more obvious goals of developing reading skills and expanding the target language vocabulary, they provided some scope for pronunciation and speaking practice. Regarding their genesis, these sequences emerged from the interaction of protectionism, competition and the affordances of learning materials with a mainly transmissive pedagogical orientation. The combined effect seemed to be that the sequence was manifested within the space defined by the transmissive constraining structure.

6.3 Traditional Grammar

Another very salient attractor in the language school was the one I designated Traditional Grammar (see also Kostoulas, 2014). Traditional Grammar sequences were most frequently encountered at the Junior and Senior programmes, where they were particularly prevalent. When I asked students to describe a typical question, using an open-ended questionnaire item, there was a surprising number of references to this instructional sequence, as seen below:

Student Quotes 15-20

15. She explains us the grammar.
16. She always explain [sic] us the grammar rules the unknown words but all of them in English.
17. She tells and examples [sic] us the grammar. We write down it to the notebook.
18. My teacher explaine [sic] as to the grammar with examples.
19. The teacher explains to me the grammar that I'm going to learn and she always says to write it down in my notebook.
20. Our teacher writes new grammar on the blackboard and she tell us to write everything down. Then she explain us [sic] the grammar and answer [sic] our questions.

The activities that made up the Traditional Grammar sequences aimed at providing learners with metalinguistic knowledge about the formal features of the English language, as well as with the ability to apply this knowledge in their linguistic output. Although there was some variety in the actual practices observed, the Traditional Grammar attractor prototypically consisted of three stages (Table 6.4): a *prompt*, a presentation (often referred to as an '*explanation*') and controlled practice activities (*practice*). These were sometimes followed up by a feedback/free production stage (*application*). This macro-structure seemed to merge two pedagogical traditions. One visible influence was the Presentation-Practice-Production model, or PPP, which connected to affordances in the learning materials. However, this appeared to be adapted in order to conform to traditional forms of language education and local pedagogical tradition, such as the 'prompting-explanation-application-consolidation' pattern described by Matsangouras (1988).

Table 6.4: Overview of a Traditional Grammar sequence

Stage	Description & comments	Timing	
Prompt	• Schematic links made between previously presented material and target structure	5'	Lesson 1
Explanation	• Teacher delivers lecture on target structure (form, use, examples)	15'-20'	Lesson 1
Practice	• Simple limited production exercises conducted individually. • Exercises corrected in plenary mode.	25'-30'	Lesson 1 (+homework)
(Application)	• Oral production of target form (Initiation, response, Feedback) • Production of short written text	5' →	Lesson 1 Homework

6.3.1 Prompting

Grammar lessons usually started with some kind of prompt, which was meant to display the linguistic structure to be taught. Sometimes, the prompt might be suggested by the textbook. Alternatively, teachers could use realia, seemingly impromptu observations ('is it going to rain?'), or tasks designed to elicit the target structure. An example of such a task is presented below:

> **Notes Extract 17**
> The class is divided into two groups (boys vs. girls), and each group is given several strips of paper. These contain fragments of a story which has been divided in strategic spots so as to draw attention to the past perfect forms.

The following extract, from one of the classroom observations, illustrates how segments from a previously taught passage were used, somewhat unsuccessfully in this instance, to prompt the presentation of a complicated language structure:

> **Notes Extract 18**
> The teacher writes the following sentence on the board:
> *Life in Amsterdam was too ... for the Franks, so they had to hide*
> and invites the students to complete the missing word. This is met with silence, so after some unsuccessful prompting, she writes in the word 'dangerous', and asks the learners why that might be the case. One student raises his hand and –seeing that there are no other offers- replies 'because is war'. The teacher adds another sentence:
> *The annexe was too small for the Franks to live.*
> She asks the learners if this was true, and some nod their agreement. The teacher seems rather frustrated now: she points out that they lived there for a number of years. '*The room was NOT too small*', and adds the word NOT to the sentence. I was left with the impression that the students were puzzled.

Prompts tended to be relatively short in length, and their purpose (as frequently stated in the lesson plans I consulted) was 'to connect the new language to what already had been taught'.

6.3.2 Explanation

The next stage of a prototypical grammar lesson consisted of a lecture in which the formal features and uses of various language structures were brought in focus. These lectures, or *explanations*, which could take up as much as half the duration of a lesson, were marked by their length and sophistication. While the teacher was presenting the grammar points, students would normally be expected to take notes on dedicated grammar notebooks. The following narrative reconstructions from my observation notes offer some insights into what this stage involved.

Notes Extract 19
The teacher announces that they are going to talk about the future tenses. She uses the IWB [Interactive Whiteboard] to project a slide with two columns, one for predictions and another one for plans. In each column she has made a list of words such as *arrangement*, *plan* etc., which are arranged in order of certainty. After explaining the model, she provides examples for each category and directs the learners' attention to the tense of the verb, checking comprehension. The learners do not seem to be experiencing much difficulty – they use meta-language fluently.

Notes Extract 20
The teacher then uses different colour pens to underline various words in the sentence, and provides the grammar rule in the abstract
TOO + adjective + TO + infinitive
She explains that this shows that the action described in the infinitive is impossible because of what is described in the adjective, writes this on the board and repeats it in Greek. Adding numbers to the board, she then asks the learners to copy this information in their notebooks (first the rule, then the examples).

Notes Extract 21
Next, Rose dictates a sentence exemplifying the Past Perfect Continuous ('*When I returned home, my daughter had been talking on the telephone for 45 minutes*'), and draws attention to the verb. She asks the learners to identify the tense and goes over its properties. This is relatively unproblematic, perhaps because it is a revision of material that has been taught at some previous lesson (and C' Class). I am struck by the eagerness of the learners to participate, and the ease with which they use meta-language (the Present Participle was identified as such almost as frequently as it was called the '-ing').

The grammar reference sections in the coursebooks were often consulted and used during the *explanation* phase. The content of these sections was often read aloud on class, and key terms were glossed over, usually but not always in English. Some teachers claimed that they were unsure about using these grammar reference sections, which they perceived to be insufficiently comprehensive or explicit. The following interview extract illustrates this belief:

Interview Extract 16
Teacher: …they present stuff very succinctly. I mean, they make it sound like 'here's what it's like, so learn it by heart', rather than 'this must be done in such-and-such a way' and 'because it has these…'
Achilleas: So how would you rather they show grammar?
Teacher: ((long pause)) Show a few more details in the rules. Because they sometimes show just 'He does; She does; It does' but they do not explain why they add '-e-s' [to the end of the word].

A similar point was raised by another teacher who also pointed out that:

> **Interview Extract 17**
> The grammar sections [in the books] were not very useful to the students, as they [i.e., the students] tend not to read them, but they are quite useful to the teachers, because they allow them to elaborate on the grammatical content of the lesson.

After prompting, the teacher explained that learners are 'of course exposed to new grammar in the language input sections of the materials', but this is not always comprehensive, as it did not cover all eventualities or exceptions to the rules. For instance, there may be a text in the students' books demonstrating the plural forms of nouns, but it's unlikely to contain enough input to cover all the irregular forms.

In order to counteract the perceived deficiency of the grammar reference sections, many teachers encouraged the use of dedicated grammar notebooks, where learners recorded the content of the lectures. Alternatively, the main points of the lecture might be provided in the form of supplementary handouts (see also Section 4.3). In one pre-observation briefing, my notes record that the teacher 'made a point of reassuring me that the students have already been given a list of words followed by gerunds and infinitives "with exceptions, verbs, adjectives etc." on which they will "of course" be tested.' When I compared the teacher's list to the grammar reference section in the coursebook, I noted that the latter contained six verbs, whereas the teacher's list spanned over half an A4-sized page. The rationale provided by the teacher for supplementing the coursebook so extensively was that she felt it was necessary in order to better prepare her learners for an upcoming term test.

The *explanation* phase, which roughly corresponded to the presentation phase in a PPP sequence, was a typical example of transmissively oriented pedagogy. This pedagogical orientation might be related to transmissive affordances in the materials, but it is interesting to note that the actual classroom practices were even more strongly influenced by traditional pedagogical practices than what the learning materials would suggest. One possible explanation involves the influence of the competition intentionality, which aimed at maximising learning outcomes (Section 5.5), and which connects to the extensiveness of grammar content with which the learners had to engage.

6.3.3 Practice

The third stage in this prototypical sequence involved controlled practice of the language forms that had just been presented. This stage had a superficial structural resemblance to the 'practice' component of PPP instructional sequences; in fact, the activities used in class often appeared to have been designed for such sequences. However, there were also notable differences. Whereas the theoretical provenance of PPP lies in the audio-lingual method (Richards & Rodgers, 2014), and it relies on habit formation as a means for developing linguistic competence, the *practice* stage as observed at the language school often served as a trigger for low-level cognitive processes. Another important difference between the *practice* stage and audio-lingual theory was the high emphasis placed on practice.

During the *practice* stage, students engaged with a number of activities that required the controlled production of language forms. Students were normally expected to complete these activities in writing, working individually, under the supervision of their teacher. Many of these activities were taken from the coursebooks, again hinting at the role of affordances, but as several students pointed out, teachers often supplemented the coursebook content with additional exercises ('our teacher gives us photocopies with exercises', 'most of them [i.e., the activities] are in our workbook and our book. The others are from the teacher.') In one of the interviews, a teacher explained to me that the exercises that were contained in coursebook were inadequate for her classroom needs:

Interview Extract 18
Teacher: …they do exercises on this. What they got, how they consolidated it.
Achilleas: Exercises which are…?
Teacher: Grammar-
Achilleas: Yes. From the book? Are they in…?
Teacher: They might be from outside the book. (5 sec) Because usually it's not… there isn't enough. There's only one or two, exercises, for each phenomenon. And these aren't enough for every student to speak.

The frequency with which the teachers supplemented the grammar content of their coursebooks seems, once again, suggestive of the influence of the competition intentionality.

Some insights into how the *practice* stage was implemented can be derived from the following extracts, from my classroom observation notes:

Notes Extract 22
Next, the students are given a handout with notes which [they] are to use in order to form sentences. Students work individually on this, and then the answers are read out in class. This is well done […] with few mistakes in the tenses…

Notes Extract 23
Following that, she presents each learner with a handout where there are several examples of sentences in future tenses. She asks the learners to identify and justify the use of each tense. This is followed by a multiple-choice exercise where the learners are asked to select between different verb forms.

Typical exercise formats included gap-and-cue items ('It has some verbs, that we have to put in the correct form'), multiple-choice exercises ('Our exercises are related to multiple choice'), or transformations ('[we] write a sentence to be the same as a sentence below'). Less typically, the practice stage also included exercises in which learners had to explicitly demonstrate metacognitive knowledge ('we do exercises like to answer questions about grammar'). This last category of exercises provides us with an example of the influence of local pedagogical traditions.

Experienced teachers might not be surprised to read that these activities were not very popular among the students. In fact, in one of the class surveys that I carried out, most students suggested that they were assigned grammar exercises 'too often'. Many complaints were also voiced in the responses to the open-ended questions, including:

Student Quotes 21–23
21. I want them to assign fewer grammar exercises because they are too long.
22. You must pay many [sic] money for pens :)
23. [Our teacher] gives us a lot of exercise to do so we will be practice, […] I don't find it helpful at all but we can't do anything about this.

However, in the classes that I observed students often carried them out with remarkable efficiency and confidence, and I was surprised to notice that they often engaged in ludic behaviour during this stage, as seen in the following extracts from classroom observation notes:

Notes Extract 24
This is followed by a fast drill, where the teacher mentioned a verb and learners took turns to provide the correct Past Perfect Continuous form. Additional practice was provided in the form of a gap-and-cue exercise from the workbook, where the learners are asked to produce the Past Perfect Continuous. After completing the exercise, the learners are instructed to exchange workbooks and correct the answers, using the key that is provided in an OHT [Overhead transparency]. The student who was sitting in front of me offered to give me her workbook if I would give her my notes. They seem to enjoy this quite a lot, and there is some mock scolding exchanged among them.

Notes Extract 25
The learners are directed to the exercises in their book (multiple matching and gap-and-cue), which they do in turns. There is some banter, as the students recite the examples: When a girl reads that 'I'm too young to drive' one of the boys points out that women shouldn't drive anyway; another subverts the assumption in his sentence by reading that 'I'm NOT too young to go on holiday on my own' to which one of his peers replies 'You are a baby'. I am rather surprised both by the facility with which the learners seem to have learnt the structure, and by their merriment.

Such instances of schismic discourse, carried out even in spite of the potentially face-threatening presence of an external observer, stood in stark contrast to the relatively sombre classroom dynamics that typified the preceding 'explanation' stage. It is possible that this behaviour hints at the high degree of confidence the students had attained, resulting from competence and familiarity. Alternatively, it may have been their way of coping with what they described as 'boring and easy' activities.

The theoretical legitimisation of the *practice* stage was provided by one of the teachers. Drawing on her extensive teaching experience, she claimed that learners tended to ignore the grammar reference sections in the textbooks, and – likewise – the grammar notes the learners made only provided a minimum of exposure. It was therefore in the practice activities where 'true learning' took place, as it was in this stage that learners 'consolidated' the language. This teacher's response offers further indication that goal of the practice activity was to further develop language awareness, rather than to foster mechanical skills formation. This observation was partly corroborated by the students. Although I had expected students to dismiss the value of mechanical grammar exercises, their answers suggested that such activities were viewed as 'the most efficient and practical way to learn something', and as a way to 'learn when we must use this form and when the other'. Intriguingly, the reason cited by the learners did not involve mechanical habit formation, as might be expected by the informing theory of behaviourism. Rather, the students' responses contained references to processes such as 'understanding', 'learning' or 'noticing', as illustrated below:

Student Quotes 24–29
25. They are helpful if you haven't understood the grammar.
26. I learn better a specifik [sic] grammar phenomenon.
27. I think they are helpful because we can see and correct our mistakes in order not to do them and improve our English.
28. They are helpful, because we can do [put?] the things that we know to action.
29. For me the exercises are like examples.

Some other benefits students associated with grammar practice activities included test preparation and general linguistic development. Some students appeared to value controlled language practice as an opportunity for self-check ('They are very helpful because I review from them if we take a test and they help me to check what I know and what I must read again'). The implication seemed to be that for in some cases at least, grammar practice activities provided opportunities for some kind of cognitive engagement with the language, and functioned as an extension of the *explanation* stage.

6.3.4 Application

Occasionally, *practice* stages might be followed by a semi-free production stage, which involved the production of oral or written discourse displaying the previously taught structures. In the lesson plans and the teachers' discourse, these stages were referred to as 'production' or 'application', suggesting the influence of both ELT and Greek pedagogical literature. The latter term in particular hints that the role of the protectionism intentionality, which valorises the technical knowledge of locally trained teachers.

Application stages were not very common in traditional grammar lessons. In fact, I did not have the opportunity to witness such an activity in any of the lessons I observed. In my experience however, and according to informal conversations held with teachers, they were generally regarded as less important than the preceding *explanation* and *practice* stages. Teachers tended to view these activities as an optional extra, which could usefully fill instruction time if the *practice* stage took less time than planned. They also suggested that it could provide feedback to the teacher regarding the effectiveness of instruction. In either case, the *application* stage did not appear to be an integral part of the learning process. Consequently, the time allocated to such activities tended to be brief. In the lesson plans that I studied, *application* stages were often allocated between five and ten minutes of instruction time, but there was a tacit understanding that these entries were included in the plan for reasons of comprehensiveness, and that they were very likely to be shortened or omitted depending on time constraints.

The activities used in the *application* stages were normally derived from the published learning materials. Such activities were often found at the end of grammar-focussed lessons, and usually involved speaking and writing for practicing the language structure that had been introduced. Although the rubrics in the coursebooks and the instructions in the teacher's books suggested that the oral activities could be done by the students working in pairs or groups, teachers often used them as prompts in Initiation-Response-Feedback exchanges, i.e., the teacher would nominate students to produce pieces of discourse in response to the prompt, and would then provide feedback regarding the accuracy of the students' output. The writing activities would most commonly be assigned as homework, in order to save classroom time. The prevailing view among teachers seemed to be that classroom time could be used more productively for grammar practice, and that students would be better able to concentrate on a writing task in their home environment. As regards their pedagogical effectiveness, written tasks were viewed with some ambivalence by the teachers, who described them as 'stilted and counterproductive' and 'useful but rather tedious'.

The way in which teachers re-interpreted the activities in the coursebook at this stage of the sequence provides us with evidence that the affordances in the learning materials could shape teaching and learning practice, but did not always determine it.

6.3.5 Emergence as a localised phenomenon

As was the case with the Reading and Vocabulary attractor, the Traditional Grammar attractor was anchored in the technical constraining structure of the state space (Table 3.1). This was due, in part, to the transmissive pedagogical orientation of the affordances present in the learning materials. Looking at the affordance landscape associated with the grammar activities (Figure 4.2), one of the most distinctive feature is a valley-shaped depression, which indicates the prevalence of transmissively-oriented affordances. Perhaps unsurprisingly, this depression ran along the Junior and Senior programmes of study, which is where the Traditional Grammar attractor was mostly observed. That having been said, the discussion above also showed that the emergence of this attractor owed much to intentionalities such as competition, since teachers tended to adapt learning materials to suit their own classroom needs and pedagogical beliefs. Similarly, the synthesis of pedagogical traditions from mainstream Greek pedagogy and the ELT literature also offers hints at the operation of the protectionism intentionality.

The end product of this interaction was the development of a distinctive form of teaching and learning, which was structurally similar to the Presentation-Production-Practice sequence, but differed in a number of ways. The most salient modifications were the following: (a) a relatively brief *prompt* stage was generated through improvisation or the design of new materials; (b) during the *explanation* stage, additional emphasis was placed on metalinguistic awareness; (c) the *practice* stage, though formally similar to that encountered in a PPP sequence, seemed to supplement the *explanation*, as it aimed to foster language awareness rather than mechanical habit formation; and (d) the free production phase was either omitted or re-conceptualised as a means to test learning.

These differences are theoretically important, because they highlight how local practice can develop bottom-up in idiosyncratic ways that are responsive to local needs. It seems easy to dismiss this local pedagogical form as an example of imperfect application of the PPP instructional sequence. In such a case, the differences between the 'pure' and 'local' forms of the sequence would have been dismissed as mere 'noise', without theoretical or empirical value. However, the complexity-informed view that is put forward in this book helps us to understand the emergence of this attractor as the outcome of interactions between intentionalities and affordances, and thus provides us with a more powerful theoretical account that can be usefully brought to bear on other contexts as well.

6.4 Process-based Writing

The third instructional sequence that we will examine in detail is Process-based Writing. The name of this attractor (and Genre-based Writing; see Section 6.1.2) derives from a distinction made by Tribble (1996) between these two types of writing instruction. I have chosen to describe this attractor for three reasons. Firstly, unlike the two attractors that were presented above, Process-based Writing seems to be anchored in the mainstream constraining structure of the state space (Table 3.1), by virtue of the communicative pedagogical orientation of the activities that it comprises. As such, it provides us with a paradigmatic counterpoint to the transmissive attractors described above. Secondly, this description offers insights into the processes of change at the language school, and the tensions associated with such change. A final reason is that this sequence brings to the forefront my insider status in this setting, and it showcases how my privileged knowledge of the context helped to supplement the empirical evidence that underpins this study.

Process-based Writing had been introduced at the language school as part of a major curricular reform that I designed and helped to implement between 2004 and 2006. The main aim of that reform had been to replace transmissive teaching methods and activities, which had typified instruction at the language school, with methods and activities that drew on Communicative Language Teaching. The writing strand of the syllabi in place at the time consisted of two types of activities. Form-focused writing activities, the first type, involved the semi-free production of recently taught structures. In the other type, Genre-based Writing, students were taught the formal features of a genre, and then worked individually to produce a text within the target genre. The Process-based Writing sequence that I introduced to replace these practices had been inspired by Seow (2002), and it emphasised collaborative learning, creativity and assessment on multiple criteria other than orthographic and grammatical accuracy. In vocabulary of complexity, it involved a re-orientation of the system from the transmissive to the communicative pedagogical position, and hence from the technical to the mainstream constraining structure of the state space (Table 3.1).

Six years later, when I was conducting fieldwork for this study, it seemed that the lasting effects of the changes that I had introduced had been uneven. Process-based Writing sequences were a regular feature in the Upper Intermediate and the Proficiency programmes, as well as the Senior (New) programme, which was being gradually phased in. On the other hand, in the Junior and Senior programmes,

Genre-based Writing continued to be the dominant form of writing instruction. This observation is borne out by the distribution of writing activities in the learning materials (Chapter 4).

While I was disappointed to see how limited the uptake of my contribution to the curriculum of the language school had been, I was quite keen to see how the Process-based Writing lessons were implemented. This proved surprisingly challenging, and brought up a number of tensions between my dual role as a researcher and former Director of Studies. For example, when I expressed the wish to observe a writing lesson during a staff meeting, several teachers countered that there was not much to be observed ('just the kids working in groups'), and some teachers voiced concerns that my presence might be disruptive. The teachers' reluctance to be observed in this instance stood in sharp contrast to their otherwise relatively welcoming attitude. At the time, I assumed that it was due to their scepticism regarding the pedagogical value of collaborative activities, and perhaps the potential loss of professional face associated with the lack of strong teacher control. In retrospectively reflecting on the experience, I further believe that some teachers may have felt uncomfortable at the prospect of being observed by me, as they implemented lessons that deviated from my occasionally prescriptive guidelines. Whatever the underlying reasons, I was unable to draw on evidence from classroom observation to reconstruct this prototypical sequence, but I counterbalanced the lack of direct observational data by drawing on lesson plans and learning materials, sensitively phrased questions, and informal conversations with teachers and students.

A distinctive feature of writing lessons was that they formed a somewhat independent strand in the syllabus. To begin with, they were allocated designated slots in the timetable once weekly or every two weeks. These lessons tended to be assigned to senior teachers, who were not necessarily the regular class teachers. In a discussion with a management figure at the language school, I was told that this organisational oddity was necessitated by the relatively high demands that Process-based Writing sequences placed in terms of lesson preparation and preparing feedback for the learners. I was told that the unusual nature of the activities was a better match to the skills of experienced teachers, and that the time-intensive nature of preparation had to be reflected in the pay scale of the teachers involved.

Table 6.5: Overview of the Process-based Writing attractor

Stage	Description & comments	Timing (*)	
Collaborative writing	• Generating content / planning • Drafting • Editing • Peer-review	45'	End of Lesson 1
Individual writing practice	• Practice draft (homework)	→	Homework
Feedback	• Group feedback • Individual feedback • Portfolio document	45'	Beginning of Lesson 2 (+homework)

*The time allocated to each activity within this stage varied depending on the lesson objectives.

The Process-based Writing attractor can be conceptualised as a cycle spanning two half-lessons and a homework session (Table 6.5). In the first half-lesson, the students would collaboratively engage in a series of tasks, which ranged from generating content and planning to providing review to their peers. Students would then be assigned a writing task, in response to which they individually produced a draft text as homework. The cycle concluded with a second half-lesson, during which various forms of feedback were provided, and a 'portfolio' version of the text was produced.

6.4.1 Collaborative writing tasks

In the first stage of the Process-based Writing attractor, learners were assigned to two to three different groups, comprising three or four students each, and worked together on a number of activities, which corresponded to different writing sub-skills, such as pre-writing activities, drafting, editing and reviewing each-others work. All, or most, of these skills were practiced in every lesson, although emphasis would shift from one skill to another depending on the objectives of the particular lesson. Table 6.6 lists some learning objectives associated with Process-based Writing, in abbreviated form.

Table 6.6: Extracts from the D' Class syllabus (writing)

Unit 1	Unit 2	Unit 3
Descriptive Writing • Selecting a topic • Paragraphing • Note-making • Editing (Content)	Emails • Structuring an email • Generating content • Using prompts • Editing (Style)	Formal transactions • Expanding • Rephrasing • Proofreading

In the pre-writing step, learners in each group would typically engage in collaborative brainstorming and planning activities, based on prompts provided by the teacher. The expectation was that these activities would be carried out in English, particularly among the more advanced learners, in line with the strict English-only policy of the language school. However, based on comments informally provided by students, it seems that linguistic resources in both Greek and English were pooled in order to complete such tasks bilingually.

Following that, students were tasked with producing different parts of a larger text. For instance, each student in a group might be assigned a different paragraph of a five-paragraph argumentative essay. Detailed step-by-step instructions would be provided in writing to each student in order to facilitate the writing process. Learning resources, such as dictionaries, were made available to foster autonomy. At this stage, the teachers' role was limited to supervising the students work, in order to ensure efficiency, and to acting as a resource when assistance was requested.

Next, the students would edit and proofread the texts produced by other group members, according to instructions provided by their teacher. These activities tended to be quite focused: on different occasions, students might be asked to focus on spelling, format, or register. Sometimes the teachers would demonstrate how to give feedback, and a practice feedback activity might be done in plenary mode, to ensure that the goal of each editing activity was universally understood, before the students engaged with each others' work. The edited drafts were then collated, and the students worked in groups to ensure textual cohesion. The result of this process was a collaboratively-produced text, sometimes referred to as the 'practice' draft.

In the last activity of this stage, learners would be asked to read the practice drafts produced by other groups and to provide summative feedback using an analytical grid. The aim of this step was to raise the students' awareness of the multiple factors that contributed to successful written communication, as well as to familiarise them with the analytical grading schemes encountered in certification examinations.

As can be deduced from the description above, a key feature of this stage was its collaborative nature, which strongly differentiated it from sequences described in the previous sections. This seemed to generate some tension at the language school, perhaps because at least some teachers felt apprehensive about relinquishing control of the learning process to the learners. The following interview extract illustrates what I believe to have been widespread scepticism regarding collaborative modes of work:

Interview Extract 19

Teacher: I understand the rationale, but to be frank I do not know how well it works. This is the first year I am working on this book, and I see it as a trial of sorts, like=
Achilleas: =an experiment?
Teacher: Yes, an experiment, because I don't want to reject something without having tried it out. And, OK, in theory it's not bad.
Achilleas: And how will you judge if this experiment has worked by the end of the year?
Teacher: ((laughs)) Why? Is anyone going to ask me?
Achilleas: If I didn't value your opinion, I wouldn't ask.
Teacher: ((laughs)) OK, if that's the case, for you. I'll see if the children are more satisfied, if they write better, if I can teach all the syllabus.
Achilleas: At this point, can you make an estimate?
Teacher: ((long pause)) I haven't noticed much difference.

Some other reservations that teachers expressed, when informally queried about their attitude towards collaborative writing, included apprehension about noise levels that these activities tended to entail, as well as the students' readiness to use Greek when not directly supervised.

On the other hand, when I asked a group of 29 Senior and Upper Intermediate students how they felt about the different components of the Process-based Writing sequence, responses were overwhelmingly positive. Activities such as brainstorming, planning and peer-correction were consistently described as 'very useful' and 'quite useful' by more than three quarters of the respondents. In fact, only one student consistently responded that all the activities were 'a waste of time'. Responses to open-ended questions in that questionnaire indicated that these teenagers' positive attitudes related to the feeling that their 'voice' was valorised in the classroom. In the words of one student, 'I enjoy very much the opportunity to discuss with my friends and say what I think about important social subjects'. Interestingly, the most negative attitudes were expressed with regard to peer correction, a finding that might be attributed to interpersonal dynamics, as well as the challenge this activity posed against the traditional power structure in a classroom. The teachers' dominant role was further affirmed by the very strongly positive attitudes students expressed regarding the importance of receiving feedback from their teacher.

The task-based nature of the activities that made up the collaborative writing stage and their focus on meaning were two features that clearly oriented them towards the communicative pedagogical position in the state space. That said, the scepticism that many teachers expressed regarding communicative work, and the persistence of the traditional power structures, evident in the belief that correction should be done by teachers, hinted at the resilience of the attractors anchored at the technical constraining structure.

6.4.2 Individual writing practice

In the second stage of a process writing sequence, learners were expected to produce a first draft of a text, based on a prompt, which students often referred to as the 'subject' or 'question' (both designations being translations of equivalent terms in L1 writing instruction). Individual writing practice was normally assigned as homework, to be carried out by the students at home or at the self-access centre.

Despite the fact that producing extended pieces of written discourse involved investment in time and effort, many students expressed very positive attitudes towards these writing tasks. In one of the questionnaires, a student claimed that writing stories was the most interesting aspect of her learning experience ('the most interesting was when we were writing some stories of our minds' [i.e. stories that we thought of ourselves]), and another one stated that he enjoyed writing tasks 'because they have unusual subjects and that makes you curious'. Elsewhere in the dataset, an advanced learner commented that:

> **Student Quote 30**
> Personally, I like the writing tasks that we're assigned, these questions!!!! They are just perfect, I like writing. […] it's the best thing for a kid, when you show them that you care about what they say!

In addition to providing an outlet for the learners' creativity, and a space for voicing for their opinion, writing tasks were also considered important for instrumental reasons. The writing prompts used in this stage were modelled after those encountered in common language certification examinations, and the instructions that were given to scaffold this process were also written with the needs of examination takers in mind. This was a point that was not lost among the learners, one of whom pragmatically remarked, 'It is difficult to know without teaching how to write reviews, reports, etc. which they are in the lower [i.e., the B2-level examination], so it's useful to practice this and find information to help you'.

What is theoretically interesting about this sequence is the clear evidence it provides regarding the effects of the certification intentionality, which became increasingly important in the Upper Intermediate programme of study and onwards. In Section 5.2, I noted that this intentionality was oriented towards the mainstream constraining structure of the state space, by virtue of the communicative pedagogical orientation of the activities with which it is associated. The co-existence of this intentionality with the communicative orientation of the activities that make up this attractor provides us with a glimpse of the connections between the intentionality and teaching practice.

6.4.3 Feedback

The final stage of a Process-based Writing sequence took place at the beginning of the next writing lesson. During this stage, learners received feedback on the draft texts they had produced, which was then used towards producing a final or 'portfolio' versions.

Feedback was provided in three forms: At its simplest, it consisted of comments written directly on the learners' script. In addition, learners would be provided with an Analytical Feedback Form containing a grade and formative comments. The analytical criteria on which the assessment of the text was based were the same as the ones used for peer-correction at the end of the *collaborative writing* stage. Lastly, feedback might be provided in plenary mode, as follows: Noteworthy extracts from the students' scripts were presented on the board, on overhead transparencies or in handouts, and students were invited to comment upon them. The process-based writing instructional sequence was completed with the production, by the students, of a 'portfolio' version of their paper, which was used for assessment.

The activities, procedures and materials associated with the *feedback* stage were intended to promote communicative language teaching. To that end, the syllabus documents and assessment rubrics emphasised the need to provide meaning-focussed feedback, and to assess the students' writing performance based on several criteria (e.g., content, cohesion, coherence and accuracy). In part, this was driven by the need to align feedback procedures with the grading criteria used in certification exams that were informed by communicative language teaching ideology (see Section 5.2). However, there was a proclivity among teachers to emphasise grammatical and orthographical accuracy in their feedback, especially in the comments they wrote directly on the students' scripts, and in plenary feedback documents, which were designed by the teachers themselves. However, and interestingly, the rigid structure of the Analytical Feedback Form may have influenced teachers to provide feedback that was more balanced. The implication seems to be that while the certification intentionality that was sustaining this attractor was not strong enough to produce a behavioural change in the teachers' practices, the cumulative effect of the intentionality and the affordances generated by the feedback form did have the desired result.

6.4.4 Reorienting the system

Unlike the attractors that were described in the previous sections, the Process-based Writing attractor was anchored on the mainstream constraining structure of the state space (Table 3.1). The communicative underpinnings of this attractor were mostly visible in the first stage, during which students engaged in collaborative meaning-focused activities. Additional traces of communicative influence were in evidence at the *feedback* stage, as the assessment criteria took into account the negotiation of meaning alongside formal features of the students' output. In addition, this instructional sequence involved a radical re-conceptualisation of the ways in which the classes operated. Classroom discourse shifted from metalinguistic instruction to what the students described as 'unusual' and 'important' topics; similarly, the learners' voice was legitimised, as they were called to bring their creativity and knowledge to bear on the production of text that was judged in terms of its content. This constituted a departure from the norm at the language school, where classroom discourse tended to be under tight teacher control.

In addition to providing a counterpoint to the transmissive attractors, the description of a communicatively informed prototypical sequence offers insights into the potential for change at the language school and the challenges associated with it. The move away from entrenched transmissive practices was, in part, a product of a rigorous and sustained curricular reform that aimed to disrupt protectionist influences at the language school, and to reorient practice from the technical constraining structure to the mainstream one. Although this reform had been only partially successful, residual effects were still evident in the routinised procedures associated with this attractor, and in the communicatively affordances generated by the learning materials used in this sequence. In addition to the residual effects of the top-down pressure that had been applied to the system, this reorientation might be associated with the phase shift in the dynamics of intentions at the onset of the Upper Intermediate programme. In Chapter 4, it was noted that as practice became more exam-oriented, the influence of protectionism, which might have counteracted the curricular reform, tended to wane.

While many students seemed positively predisposed towards this change, there is evidence in the data to suggest a certain degree of tension between the communicative orientation of this attractor and the intentionalities that prevailed elsewhere at the language school. For example, many teachers appeared to hold ambivalent feelings towards Process-based Writing sequences, as might be inferred from the apprehension they indicated when I requested permission to observe them. Additionally, there appears to have been some tension between the way Process-based Writing lessons were conceptualised in the syllabus and the way they were implemented in practice. An example of this tension was found in the *feedback* stage: although instructions in the syllabus tended to stress the importance of assessing the students' work on the merit of multiple criteria, it appeared that some teachers were more likely to give feedback that emphasised grammatical and orthographic accuracy.

From a theoretical standpoint, the discussion of this attractor offers us glimpses into the complexity of the interactions that take place within a complex system such as a language school. I concluded Chapter 5 by describing how multiple intentionalities that are present in a system form 'dynamics of intentions' , and I hinted at ways in which the balance among these interacting intentionalities kept changing – from lesson to lesson, from level to level and from one year to the next. The existence of an attractor that was anchored in the mainstream constraining structure hints at a localised phenomenon, in which the certification intentionality became strong enough to re-orient the system, at least in one syllabus strand and in some levels of instruction. This, in turn, offers clues about the ways in which systems can evolve.

6.5 An evolving system

The prototypical sequences that were described above are only a small part of all the teaching and learning activity that took place at the language school. Even from such a partial reconstruction, however, it is possible to make some tentative observations about the system as a whole.

I concluded Chapter 4 by noting that the affordances in the learning materials showed traces of three processes of change. This observation seems to be mirrored in descriptions of the attractors. For example, although many of the attractors in the system were anchored in the technical constraining structure of the state space (Table 6.1), there were also a number of 'pockets' in the curriculum of the language school that were anchored on the mainstream constraining structure. These included the Speaking and Listening strand of the syllabus, as well as parts of the grammar and writing strands, and they appeared to correspond to what I defined as spatial differentiation in the affordance landscape (Table 4.7). Similarly, the complementary distribution of transmissive and communicative attractors is consistent with the phase shift observed in the affordance landscapes (Table 4.8). In the Junior and Senior programmes of study, the attractors were almost exclusively anchored on the technical constraining structure, but from the Upper Intermediate programme onwards, I noted the existence of communicatively oriented attractors, such as Inductive Grammar, Process-based Writing, and Integrated Oral Skills. Finally, there is also some faint evidence of diachronic change. Table 4.9 shows that the activities in newly introduced materials tended to generate communicatively-oriented affordances. This is mirrored by the prevalence of attractors like Process-based Writing, in the programmes of study where these materials were used. These parallels between affordances and actual teaching and learning show the interconnections between the two constructs, and hint at the theoretical utility of affordances in the description of complex systems.

It should be noted, however, that these change processes were neither uniform nor monotonic. With regard to the oral skills, for example, communicative speaking tasks were introduced at the exam-oriented courses, but listening tasks were highly transmissive, possibly due to the washback effect of the same examinations. Similarly, in the redesigned Senior (New) programme the shift towards Process-based Writing coincided a stronger emphasis on traditional grammar, i.e., there was a bifurcation of instructional patterns. In addition, the effects of the changes were differentiated due to the teachers' mediation, as evidenced by their proclivity to provide grammar-focused feedback even in the Process-based Writing sequences. These unexpected findings highlight the complexities involved in affecting change on an educational setting, and put linear assumptions of causality to the test.

A second comment that can be made regarding the attractors relates to the role of intentionalities in animating the system. The observation was made several times in this chapter that teaching and learning activity was not produced by the unthinking application of the action possibilities inherent in the learning materials. Rather, intentionalities such as protectionism and competition seemed to exert a strong influence on teaching and learning. For instance, in the discussion of the Traditional Grammar attractor, I noted that the transmissive pedagogical orientation implicit in the learning materials was amplified by transmissively oriented intentionalities, such as competition and protectionism. By contrast, in the discussion of the Process-based Writing attractor, the affordances implicit in the materials were tampered by scepticism about communicative language teaching; however, the confluence of certification and communicative affordances could have lasting effects, as was the case with the Analytical Feedback Form. On the whole, the attractors should be understood as products of the interaction between intentionalities and affordances, which can sometimes have unexpected results. This finding adds nuance to our understanding of the Greek ELT context, which has tended to interpret practice in linear terms as a product of credentialism (e.g., Papaefthymiou-Lytra, 2012), or as an outcome of uncritical 'consumption' of coursebook material (e.g., M. Papageorgiou, 2002).

This chapter looked into the shape of teaching and learning in the language school. This was described using the construct of attractors, defined here as preferred modes of teaching and learning, or prototypical instructional sequences. Seven attractors were outlined, of which three were described in detail, and it was noted that these were anchored either on the technical or on the mainstream constraining structure of the state space. The attractors were described as products of the interaction between affordances in the learning materials and intentionalities (or, perhaps more accurately, dynamics of intentions). The genesis of these attractors was explained as a process of emergence, and it was seen how such emergent processes can give rise to localised phenomena. This discussion brings together all the aspects of the system that were defined in Chapters 3, 4, and 5, by showing their interconnections. Having made this point, it is now time to take a step back and make a holistic appraisal of the complexity perspective that was used in this book, and this will be the focus of the final chapter.

7 Using complexity to describe a language school

In chapters 3 to 6, I showed how a language school can be described through a complex systems perspective. The building blocks of this description were not the elements that make up the school, such as teachers, learners, classrooms, books. Rather, the school was described from a more holistic perspective, focussing on four themes: the state space of the system (Chapter 3), the affordance landscape(s) that were generated from learning resources (Chapter 4), the intentionalities that drove the system (Chapter 5), and the regular patterns of teaching and learning, or attractors, that emerged in the system (Chapter 6). In the final chapter of this book, these threads will be tied together, and in the process, the descriptive and theoretical potential of this perspective will be outlined.

7.1 Situating a system in place and time

What I hope that this study has demonstrated is that CST can be an epistemologically fruitful and theoretically productive frame for describing phenomena in ELT. Some evidence of this can be seen even from the outset of the description. The first step in a complexity-informed study is the definition of the phenomenon that interests us, a process that involves specifying its boundaries and determining the timescale that is most useful for our purposes.

7.1.1 Rethinking boundaries

A complexity-informed description of a system, such as a school, begins by defining its boundaries, a process sometimes called framing the system (Byrne, 2009). This is not a straightforward task, because the elements that make up a complex system are densely interconnected, and it is hard to tell where the boundary might lie. This challenge is, of course, not exclusive to complexity-driven empirical work, and rather than reinvent the wheel, it seems reasonable to look to other disciplines for epistemological guidance. Indeed, similar issues are raised in the research methodology literature about ethnography (Agar, 2004) and case study research in general (Yin, 2014). For example, Yin (2014) emphasises that 'a case study is an empirical inquiry that investigates a contemporary phenomenon in its real life context, especially *when the boundaries between the phenomenon and the context are not clearly evident*' (p. 15, emphasis added). Similarly, Miles and Huberman

(1994) caution researchers that the boundaries of cases are 'not quite as solid as a rationalist might hope' (p. 27).

This is, of course, sound advice, but there are two points with which I would like to take issue. The first point is purely practical: while research manuals do alert researchers to the fuzziness of boundaries, they stop short of specifying how such boundaries might be drawn if we wish to use criteria other than pragmatic considerations. This is a potentially problematic situation for research that takes an analytical perspective, as it can be hard to theoretically justify which elements were included in a description and which ones were left out. The complexity perspective I have taken in this study offers us an epistemologically satisfactory solution to this challenge. In complexity-informed studies, a precise demarcation of the system is of secondary importance, because our interest does not lie on individual system constituents or on their activity. The description of the language school, for instance, focussed on observed collective phenomena, and on the processes from which they emerged. This being the aim, a comprehensive inventory of which elements constitute the system is really besides the point, and a precisely specified boundary seems relatively less useful.

The second problem associated with boundary-setting is ontological. In the quotes by Yin (2014) and Miles and Huberman (1994) that were cited above, and in similar discourse, the linguistic choices seem to imply that boundaries do exist in a realistic sense, and that the challenge facing researchers is how to discover where the disconnect between the system and the environment lies. My concern with such an outlook is that by artificially isolating the school from its environment for the purposes of inquiry, we run the risk of losing sight of interesting interconnections that develop between the system and its context (Cilliers, 1998). Such an analytical outlook is sometimes necessary, and indeed, it is one of the cornerstones of the 'scientific method' as defined by the scholars of the Enlightenment. But as I explained in Section 2.2, it can also be reductive and unsuited to the study of situated phenomena, as it does not account for the contextual influences.

Complexity-informed work departs from this 'reductive premise' by reconceptualising the very notion of boundaries. Rather than thinking of the 'edge' of the system as the point that separates it from its surroundings, we can think of it as an interface that connects the system to the structures in which it is embedded (Juarrero, 1999). Even though the boundary / interface remains loosely specified, or 'fuzzy', we can still describe the interactions between the components on either side, endeavour to account for why and how they arise, and trace what their effects might be. Thinking of the language school in this particular study, in Section 5.6 we noted that it tended to seal out certain external influences, effectively isolating itself from their influence; and yet, some aspects of the school's activity, such as the Cultural Awareness intentionality, evidenced traces of outside influence (Section 5.4). This apparent contradiction is theoretically interesting, because it hints at the tendency of the school to resist change (cf. Kostoulas, 2014). In other words, a complexity-informed understanding of boundaries as semi-porous structures (or even selectively porous ones) can yield insights into the quality of the interconnections in and around the system, and can therefore be theoretically productive when it comes to describing situated phenomena.

7.1.2 Thinking about timescales

Framing a system in terms of its constituent components is only one aspect of its definition. The other aspect is what we can call temporal framing, i.e., deciding which timescales we are interested in observing. In this case too, CST has theoretically generative potential, because it can help us to understand how dynamical processes that occur in different timescales interact with each other.

In Section 2.3.2, I noted that complex systems are enmeshed with other structures of bigger or smaller magnitude, which can themselves be conceptualised as overarching or embedded complex systems. Although Byrne and Callaghan (2014) convincingly argue that these systems are overlapping rather than hierarchically organised, for the purposes of this particular discussion and in the interest of simplicity, we can choose to view them as a hierarchy of smaller systems nested within progressively larger ones, not unlike Russian dolls. So, for instance, a learning activity may be thought of as being embedded in lesson, which is in turn embedded in a syllabus, a multi-year curriculum, as well as proximal (e.g., national) and global ELT structures (Figure 8.1).

Davis and Sumara (2006) point out that the dynamics that develop within these systems operate at different timescales: lower-order systems, such as language lessons, are more volatile, whereas higher-order systems are relatively more stable, as might be the case for the linguistic, pedagogical and political considerations underpinning mainstream ELT cultures. In fact, when our focus is on lower-phenomena, such as a language lesson, it might even be descriptively expedient to treat higher-order phenomena as static structures. However, it is theoretically crucial to remember that the difference between higher- and lower-order systems is one of degree, as it relates to the different rates of change, and not one of quality. This means that the same theoretical perspective can be brought to bear on descriptions focussing the micro- and macro-level, and this is one main appeal of CST.

Figure 8.1: Embeddedness of systems (based on Davis and Sumara, 2006, p. 27)

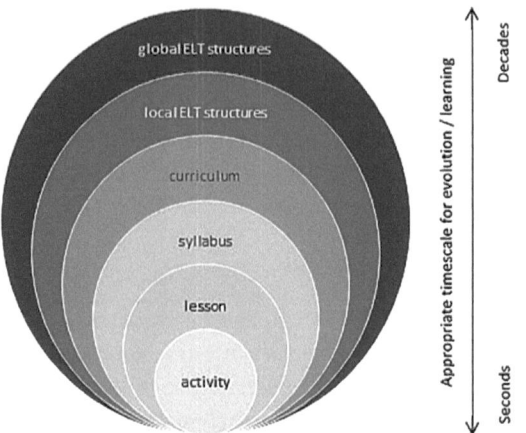

These theoretical arguments were empirically borne out in the study of the language school, since the dynamical processes that I documented seemed to be associated with different rates of change. Lessons were relatively volatile, as the pedagogical orientation of their constituent activities could change multiple times within a 60- or 90-minute time frame. Viewed from this level of activity, the syllabus and the curriculum appeared as static structures, and they appeared to constrain the content and format of individual lessons. However, when 'jumping levels' (Davis & Sumara, 2006, p. 107), processes of change become evident in the curriculum level as well, as the introduction of a new syllabuses every few years was influenced by shifting dynamics of intentions. Notably, while the rates of change were different, the drivers that sustained it (i.e., the intentionalities) were the same, and this observation helps us to connect phenomena on different timescales.

What a complexity-informed conceptualisation of ELT phenomena offers us, then, is a theoretically productive way of thinking about process and structure, as well as the dynamic interactions between them. It helps us to understand structure as a contingent manifestation of emergent processes, which constrains volatile lower-order phenomena, and which is also shaped by them. From an epistemological perspective, this implies that there is important common ground between the disciplines that have focussed on different levels and timescales, such as SLA and psycholinguistics, education theory, the sociology of education and the history of ELT. It also presents the opportunity for developing a unifying meta-discourse that connects these different perspectives.

7.2 Looking at the structure of the system

When describing phenomena in ELT from a complexity perspective, looking at the structure of the system can yield useful theoretical insights, some of which I will list in the pages that follow. But before proceeding, it is important to reiterate a point made in the previous section: the distinction between the structure and the activity of a system is a product of our epistemological perspective, and what appears as structure in one timescale can be conceptualised as activity at a different level. With this caveat in mind, the structure of the system can be conceptualised as the collective set of constraints and action possibilities (affordances) from which the activity of the system emerges. These constraints and affordances can be features of the system itself, or they might relate to the structure of superordinate systems in which the system is embedded. In the complexity literature, these are often described by means of topographical metaphors, such as state or phase spaces (Larsen-Freeman & Cameron, 2008), fitness landscapes (Byrne, 1998) or ontogenetic landscapes (Thelen & Smith, 1994). In these depictions, features such as depressions, valleys, ridges ('separatrices', to be more technical) and saddle points represent differences in the frequency with which a system finds itself in a particular state, or properties of the system that privilege certain states.

In Chapter 3, I described how the professional literature could loosely constrain the activity of the language school by specifying a set of 'legitimate' combinations of linguistic, pedagogical and political beliefs that underpin language teaching and learning. These were summarised in Table 3.1, as the technical, mainstream and critical constraining structures in the system's state space. The other aspect of the system's structure that I described was the affordances that were created by the learning resources (Chapter 4). To illustrate the role of the learning materials, I used what I called an 'affordance' landscape, which showed the likelihood that certain pedagogical forms materialised. This description of the system structure enabled the generation of several observations about the school and the setting in which it was embedded. These are presented below, and followed by a discussion of the possibilities afforded by CST.

7.2.1 Some tentative findings…

When describing the constraints and affordances in the language school, one of the most salient observations is that these seem to privilege two types of teaching and learning: a traditional, transmissive mode of instruction and one informed by the communicative approach. This was first mentioned in description of the state space (Chapter 3), where I noted that overarching structures such as local pedagogical cultures and the mainstream cultures of global ELT privileged transmissive and communicative pedagogy, respectively (Table 3.1). A similar pattern was also seen in Chapter 4, where we looked at the learning materials and the affordance landscapes that they generated, which were informed by transmissive or communicative pedagogy. Taking into account the pragmatic difficulties associated with materials generation or modification, these affordances tended to make these two forms of language learning and teaching likelier than others.

While one needs to be cautious of over-generalising, the salience of these two pedagogical modes in the system's affordance landscape hints at three significant insights. Firstly, it provides scope for refining conventional wisdom about ELT in Greece, which holds that learning materials are predominantly concerned with transmissive grammar instruction (Kostoulas, 2007; Prodromou & Mishen, 2008). This tendency was partially confirmed, but the systematic study of the learning materials in the language school seems to suggest rich affordances and dynamism that existing accounts risk concealing. Secondly, the dynamics between the two attractors hint at tensions between the mainstream and technical paradigms. More broadly, they may be seen as evidence both of the encroachment, into the local context, of the ELT ideology prevailing in the Anglophone West, and of the resistance stemming from local beliefs and practices. Thirdly, it is interesting to note that the critical paradigm was not represented in the learning materials, even though there was evidence in the data of a critically-oriented intentionality (International Integration, Section 5.3). This is suggestive of the power relations that ran through the language school. Since this critically-oriented intentionality was primarily associated with the students, it would appear that the students' needs were more likely to be catered for when they aligned to the beliefs and interests of other agents in the system.

The study of the language school also revealed that the relative strength of the two pedagogical modes were influenced by three parameters: the type of lesson, the level of instruction, and diachronic change. Activities focussing on grammar, vocabulary and reading tended to be transmissive, whereas activities aiming to develop the oral skills and writing were more likely to be communicative. In addition, the transmissive mode of teaching and learning was considerably stronger in the early stages of instruction, rather than in the exam preparation courses in which more advanced learners were enrolled. Finally, it seemed that the transmissive mode waned, and the communicative one was correspondingly strengthened, as the language school evolved over time. Caution should be exercised before claiming that such a re-orientation is part of a broader change in ELT in the Greek context (or the periphery of the Anglophone world, for that matter), but it is an intriguing possibility, which may be further explored by similar work in other settings. In addition, it would be interesting to try to confirm this finding with longitudinal or follow-up research, for which the insights developed in this study might serve as starting points.

7.2.2 ...and some theoretical insights

Taking a more theoretical perspective, the complexity-informed description of the system's structure led to a number of insights about how activity emerges in the system, about the dynamism of the system as a whole, and about the way in which the present structure of the system connects to its past and future.

In the previous section, I noted that the system's structure privileged two modes of instruction. It would seem that there was overlap between the transmissive and communicative depressions in the affordance landscape on the one hand, and – on the other – the technical and mainstream state-space constraining structures, which were hypothesised to constrain the school's activity. The overlap between the affordance landscapes in the language school and the constraints generated by local pedagogical traditions and the global cultures of ELT hints at the ways in which lower- and higher-order complex systems may interact. This is a pattern that will be discussed further in connection to intentionalities and attractors, but for now I will limit myself to the remark that the affordance landscape can help to show how constraining top-down processes (such as the views recorded in the professional discourse and literature) relate to localised phenomena, such as the ones observed at the school.

A focused look into the affordance landscape of the language school also helps us to understand that the system's structure is dynamic, an observation that relates to the points raised previously about timescales and rates of change (Section 7.1.2). By comparing the affordance landscapes generated by various activity types, I noted that activity types were connected to different sets of affordances: speaking activities, for instance, tended to generate stronger communicative affordances than grammar activities, which were associated with transmissive pedagogy. It is easy to see how variation could lead to shifts in the affordance landscape within the space of a single lesson. Similar patterns of change, operating at a different rate, were observed in the diachronic development of the school, as well as in the learners' progression from beginner to advanced programmes of study (Section 4.1.7). From an epistemological perspective, the implication is that the holistic description of the school need not be monolithic and undifferentiated; rather, the dynamic conceptualisation of structure allows for a more nuanced discussion of the phenomena that emerge from it.

The study of the system's structure is also useful for a third reason. Because complex systems are autopoetic (i.e., they shape themselves) (Maturana & Varela, 1980), their present structure is a product of their past structures. This means that a systematic study of the affordances that are currently present in a language school can be used to shed light into its past. In this study, it was assumed that whenever a syllabus or a set of learning materials was created, this act of creation gave material shape to the dynamics of intention that were prevalent in the system at that particular time. In fact, one may even view the learning materials as a 'derived intentionality' (Stelma et al., 2015), a kind of intentionality that has been sedimented in an artefact, and is in a sense suspended in time. Understanding the connections between the past and present states of the system means that even a synchronic research design can yield insights into the system's historicity (Dörnyei, 2014; Juarrero, 1999, Larsen-Freeman & Cameron, 2008).

7.3 Examining the intentionalities that drive the system

In Chapter 5, I looked into the intentionalities that emerged in the language school, which were defined as the forces that drove and sustained the activity in the system (Kostoulas & Stelma, 2017). Predominant among them was a preoccupation with proving English language proficiency through certification. I also documented a desire, among students, to integrate in transnational discourse communities, which achieved coherence through the English-mediated communication. A third intentionality involved providing learners with information about the English-speaking culture, which was restrictively equated with the mainstream culture of the British Isles. Fourth, there was competition with the state education system. Finally, there was empirical evidence of a protectionist intentionality, i.e., a concern to protect the professional interests of the local ELT professional community.

The understanding of intentionality that is advanced in this book, as well as publications like Stelma et al. (2015) and Kostoulas & Stelma (2016), was shaped inductively by the needs of this study. Its conceptual specification was driven by the need to connect the structure of the system to its activity, by showing how aspects of system structure come together to produce activity, and how activity then recursively shapes the system. This conceptual work is presented below and followed by a discussion of how the construct of intentionality helps to advance our understanding of the system.

7.3.1 Defining intentionality

The conceptual definition of the intentional drivers (intentionalities) was derived from the literature (Stelma, 2011, 2013; Tudor, 2001; Young et al., 2002) and engagement with the data. A starting point for this discussion is Tudor (2001), who suggests that activity in an educational setting, such as a school, can be understood with reference to what he describes as 'rationalities', i.e., the 'perceptions and goal structures' that drive language education (2001, p. 33). While this terminological choice connotes conscious rational reflection and deliberate choice, Tudor argues that:

> [t]hese rationalities may not be explicitly formulated in the way we find in writing on methodology, or in official policy statements, but nevertheless influence the meaning which classroom learning activities assume. (2001, pp. 33–34)

Tudor (2001) lists five groups of rationalities as starting points for empirical inquiry, namely student rationalities, methodological rationalities, socio-cultural rationalities, institutional and corporate rationalities, and teacher rationalities. He goes on to note that these groupings are not always cohesive (i.e., there may be several rationalities present in each group), their influence is unlikely to be symmetrical, and some may present themselves more readily for empirical investigation than others. While bearing in mind these caveats, Tudor (2001)'s taxonomy is analytically helpful in at least three ways: (a) it draws attention to the plurality of drivers that are present within a system; (b) it provides us with a theoretical construct for conceptualising the collective activity of multiple and possibly heterogeneous agents; and (c) it brings to the forefront the role of shaping influences that stem from outside the system.

A second strand of scholarship that contributed to the development of intentionality came from ecological psychology. Young et al. (2002) provide us with a way of conceptualising the relation between intentionalities and the environment in which they arise, which was viewed as a hierarchical set of constraints. The intentionalities that emerge within these constraints come together to form overarching 'dynamics of intentions'. This conceptualisation was further elaborated by Stelma (2011, 2013), who explained how intentionalities emerge from the resources and expectations that are present in a system. An important way in which Stelma's model advances the work of Young et al. is that he showed how activity driven by intentionality can add to the system's degrees of freedom, and thus intentionality-driven activity can transcend the constraints in which the intentionality was created. This is a crucial distinction, because it suggests that intentionality is not merely a product of the system's structure, but can also have generative effects over it.

The description of intentionality that was put forward in this book builds on these contributions, although my focus is on describing the features of intentionality, which I derived inductively from the study of the language school. Intentionalities, I argue, are collective entities, which sometimes synthesise apparently incompatible beliefs, as was the case with protectionism (Section 5.6.4). Intentionalities are mostly shaped by local conditions, as was seen in the discussion of the cultural awareness intentionality (Section 5.4.3). As also noted by Stelma (2011, 2013), intentionalities are generative phenomena, and are associated with specific pedagogical effects. Certification, for instance, tended to be associated with mainstream ELT ideology (CLT), whereas protectionism was associated with pedagogical and linguistic beliefs that align more closely with Greek education. However, such effects are not deterministic (Section 5.2.3), and their influence is sometimes unpredictable (Section 5.5.4). Finally, drawing on Young et al. (2002), I argued that these intentionalities come together to form broader 'dynamics of intentions' , which collectively influence the behaviour of the system.

The discussion of intentionalities facilitated the articulation of two findings about the language school, and (more tentatively) about ELT. First, it highlighted the existence of multiple heterogeneous drivers that are present in educational settings (cf. Tudor, 2001; Tudor, 2003), and in doing so, it called into question the validity of reductive attempts to explain similar phenomena. In addition, it brought to the forefront the existence of 'silent' intentionalities, which may have a visible impact in the system's activity. This finding shows the potential of complexity-informed descriptions to reveal hidden processes as well as make visible the mechanisms that suppress them.

7.3.2 Insights associated with the study of intentionalities

By empirically identifying several intentionalities in the language school, this study put forward a nuanced conceptualisation of ELT as a complex phenomenon. This is theoretically helpful for three reasons: first, it enables us to move from monocausal descriptions of ELT to more nuanced understandings; secondly, it raises awareness of hidden processes that nevertheless have visible effects; and finally, it highlights the unspoken power structures underpinning the shape of teaching and learning.

With regard to the first insight, the study of how intentionalities interact add to our understanding of ELT in Greece, which is sometimes reductively described as being exclusively concerned with certification (e.g., Angouri et al.,

2010; Papaefthymiou-Lytra, 2012). Furthermore, a causal account that is based on dynamics of intentions (i.e., the synthesis of intentionalities) provides a convincing explanation about an apparent paradox, which scholarship on Greek ELT often ignores or stops short of resolving: Given the obvious value attached to the (predominantly communicative) certification examinations, why has their washback effect tended to be resisted in pedagogy (see e.g., Kostoulas, 2014)? The usual caveats about projecting the findings of a single case study onto a broader setting (Greek ELT in general) obviously apply, and it would be interesting to investigate whether traces of such intentionalities are present outside the language school and what forms they might take. The initial probes into specific intentionalities that were made by this study could provide the analytical categories for such confirmatory investigations, and the conceptual insights about intentionality might serve as a theoretical foundation to inform such a research agenda.

A second theoretically interesting insight relates to the role of intentionalities that are not readily observable. In the preceding chapters, it was suggested that many aspects of the language school's operation could be interpreted in the light of interaction between Certification and Protectionism. Of these, Certification was salient both in the literature about ELT in Greece and in the discourse of the language school (Section 5.2). Protectionism (Section 5.6), on the other hand, seemed to function as a 'silent' intentionality. Even though the shaping influences from which it emerged were traceable in the data, there was little evidence of the intentionality as such in the participants' discourse, and I was unable to locate references to protectionism in the ELT literature. This paradoxical observation demonstrates the potential of the intentionality construct (and, more broadly, CST) for generating economical and theoretically coherent accounts of a system's behaviour, by synthesising apparently disparate shaping influences. What this study did not do, however, is attempt to explain why Protectionism is absent from the discourse, a question to which Critical Discourse Analysis (Fairclough, 1995; Pennycook, 2001) are likely to yield intriguing responses.

A final point of interest relates to the differentiated impact of intentionalities. While Certification and Protectionism had a visible effect on both the learning materials and the pedagogical practices of the language school, the students' desire to integrate to transnational discourse communities (International Integration, Section 5.3), did not seem to have a similar impact. This finding hints at asymmetries in the power structures that ran through the language school. Regrettably, the data at hand did not support an in-depth investigation of such differences, but such an inquiry would likely yield interesting insights about the power structures in which ELT is embedded, and it therefore holds promise for research informed by the critical education tradition.

All of the insights that were listed above hint at the challenges associated with making linear connections between specific intentionalities and effects. Put differently, it seems uncertain that isolating any intentionality and attempting to relate it to pedagogical practices would be a productive way forward. Rather, what this study has demonstrated is that a holistic outlook, which looks at the dynamics of intentions and relates them to the entire system, appears to offer greater descriptive and theoretical potential.

7.4 Pedagogical activity emerging in the system

The final set of findings that the complexity-informed perspective of this study made possible pertains to the actual pedagogical practices that took place at the language school. After succinctly going over the seven prototypical instructional sequences that were defined in this study, I argue that they were products of the interplay between intentionalities and learning resources. I also suggest that they indicate a heightened role of the local context in shaping practice, and finally I develop a hypothesis regarding the underlying mechanisms that connect the language school to processes of globalisation.

7.4.1 Pedagogical activity as an outcome of intentionalities and resources

As expected from the descriptions of learning resources and intentionalities, the prototypical sequences observed at the language school were either predominantly transmissive or predominantly communicative (Chapter 6). Among the former, I found evidence of: (a) transmissive grammar teaching, seemingly derived from the grammar-translation method and presentation-practice-production sequences; (b) sequences consisting of intensive reading activities and of vocabulary teaching that relied on memorisation; (c) genre-based writing, which aimed at mastering and reproducing the formal features of specific text types; and (d) 'review' sequences, which featured recall and mechanical practice of grammatical and lexical knowledge. Communicative sequences, on the other hand, included: (a) series of process-based writing activities, (b) speaking and writing tasks, and (c) inductive engagement with grammar structures. The transmissive and communicative orientations of these sequences appeared to align with observations made in previous sections regarding the prevalence of the technical and mainstream paradigms, and the underrepresentation of the critical paradigm.

The development of transmissive and communicative pedagogical sequences is consistent with theoretical understandings of complex phenomena. For instance, it has been argued that intentionality is bound by the system's structure (Young et al., 2002), and from that it follows that the pedagogical sequences might be seen as being constrained by the transmissive and pedagogical attractors described in Chapter 4. In Stelma's less deterministic perspective, intentionally-driven activity is generated through the interplay between expectations and resources (Stelma, 2011, 2013; Stelma & Fay, 2014), which also seems consistent with the findings of this study.

Similarly, Juarrero argues the current state of the system can be used as the starting point for a hermeneutical narrative, which can disentangle 'the specific path the [system's] behaviour took' to reach its current state, and can also explain what aspects of this behaviour were specifically constrained by intentionality, what aspects were influenced by the 'by the lay of the land' and what aspects were shaped by 'the external structure of the intention's control loop' and other dynamics (Juarrero, 1999, p. 231). Juarrero's insistence on identifying the 'specific' contributions of agentic behaviour and 'the lay of the land' (the structure within and around a system), among others, seems to be at odds with the understanding of causality that I am advancing here, but we are – I think – in essential agreement that the end states of a system are jointly influenced by intentionalities and structure. We also appear to be in agreement that retrodictive approaches can help us to understand a system's structure and activity.

Overall, a theoretical perspective that sees teaching and learning activity in the language school as a product of the interaction between intentionality and resources satisfies both theoretical validity (Maxwell, 2002) and the need for fit with data. Moreover, such a nuanced account of causality seems useful in understanding patterns of activity that evidence regularity, though not absolute predictability, as is typical of education.

7.4.2 The role of local context

A second finding relating to pedagogical activity pertains to the prominence of local contextual influences. For example, the Traditional Grammar sequence (Section 6.3) was related to language ideology in Greece, and to transmissive pedagogical practices commonly encountered in Greek education (Section 2.2.1). This interpretation offers three useful insights. Firstly, it helps to explain the survival of what might appear as dated practices (when appraised against mainstream ELT professional discourses), or the rationale underpinning the combination of activities which derive their provenance from arguably incompatible ELT methods (e.g., translations and structuralist drilling). Secondly, the use of inductive research methods helped to describe the Reading and Vocabulary sequence (Section 6.2) in terms that reflected parallels with mainstream Greek education, and were a closer match to the data than common instructional sequences encountered in the ELT literature. Thirdly, the nominally discouraged yet pervasive use of bilingual wordlists offers one more example of how proximal influences (in this case, the L1 shared by most teachers and learners) impacted the teaching and learning practices of the language school. In cases such as these, complexity provides both the warrant for extending the investigation into the surroundings of the system in search of significant proximal influences, and theoretically useful interpretations of the interface between the system and its surroundings.

By relating the activity of the school to local pedagogical practices, this study addresses a significant, if unacknowledged, shortcoming of research into Greek ELT. Historical accounts of ELT in Greece, such as Soulioti (2007), have described the diachronic development of ELT as a turnover of methods and approaches that stem from the Inner Circle and spread outwards. Consequently, they look to narratives such as Richards and Rodgers (2014) and Stern (1983) as a source of analytical categories, despite the fact that such descriptions do not, indeed cannot, take local pedagogical traditions into account. Moreover, scholarship and research about ELT in Greece sometimes uncritically assumes isomorphic correspondences between the theoretical description of constructs such as 'communicative language teaching' in the literature, their instantiations in British, Antipodean and North American settings, and the way in which they are implemented in the periphery of the English-using world.

While such a perspective offers advantages of narrative convenience, it can be politically problematic and epistemologically misleading. On political grounds, there is a case for challenging accounts that invisiblise local educational traditions in favour of an Anglocentric outlook. Epistemologically, the interpretative validity (Maxwell, 2002) of such accounts is often hard to discern, because they appear to disregard local beliefs about language and pedagogy, which have potentially significant effects on practice. For such reasons, the development of descriptive accounts of situated practice seems to constitute a research priority, and complexity-informed narratives appear well suited to that goal.

7.4.3 ELT and globalisation

One final insight of this study pertains to the ways in which ELT develops against a backdrop of globalisation. There is a strand in the literature suggesting unidirectional flows of linguistic norms, cultural values and teaching methodology from the Inner Circle outwards (Kumaravadivelu, 2006a; Phillipson, 1992, 2004). However, the empirical data from this study do not fully confirm such a thesis. To begin with, there was a perhaps surprising absence of Anglocentric cultural content in the learning materials (Kostoulas, 2015a). Moreover, there appeared to be active resistance to methodological intrusions into the local context (Section 5.6), and there were some indications that when Inner Circle cultural values and linguistic norms were invoked, this was done for the benefit of the local ELT establishment (Section 5.4). Of course, the scope of this inquiry is such that it would be injudicious to challenge the overall validity of the claims regarding linguistic and cultural imperialism, but in light of the evidence at hand it would appear that the mechanisms at work are perhaps more elaborate than what has sometimes been suggested. In the paragraphs that follow, I develop a hypothesis regarding one of these mechanisms, drawing on the data from the study.

The relation between local and globalising forces in the language school might be interpreted as a situation in which a seemingly stable state gives way to a radically different state: specifically, one in which practices associated with local pedagogy, are suddenly replaced by practices that are oriented towards the mainstream paradigm. Taking into account the drastic nature and abruptness of this change, I described it as a 'phase shift', a sudden change when the dynamics that sustained one state are 'transformed into qualitatively different dynamics governing a different space of possibilities' (Juarrero, 1999, p. 233). Juarrero argues that phase shifts are crucial in understanding the behaviour of complex systems, and goes on to suggest that a theoretical explanation of such a reconfiguration should account for three factors: (a) the internal processes and contextual influences that drove the system away from equilibrium, thus enabling such a drastic change possible; (b) the actual perturbation that triggered the phase shift, and (c) the reasons why the system took a specific new form rather than others. The exploratory nature of this particular study precluded the generation of definitive answers to the questions that Juarrero sets. However, based on the available evidence, we may legitimately hypothesise that the phase shift was associated with the valuation of certification.

In Chapter 6, it was seen that instruction at the language school was initially aligned to the technical paradigm, but this should not be taken as evidence of the absence of influences associated with the mainstream CLT. The relatively stable 'preferred' state of the system, which might be described as a state of 'dynamic stability' (Larsen-Freeman & Cameron, 2008), was sustained by complex dynamics of intentions. The underlying dynamics were typified by the confluence of several intentionalities, which served to orient the system towards the technical paradigm. However, the certification intentionality, which was associated with the mainstream paradigm, was still present in them, its effect being cancelled out by competing intentionalities. This intentionality had the potential to drive the system away from equilibrium, when the conditions were conducive to such a change.

In the literature, it is stated that phase shifts are often triggered by a change in a critical variable in the system (a 'control parameter') (Larsen-Freeman & Cameron, 2008). In such cases, when the control parameter passes a certain 'tipping point', its effects on the system tend to be disproportionately large. In the case of the language school, the control parameter appeared to be the imminence of certification examinations. As long as the examinations were a distant possibility in the students' future, the system tended to operate in its preferred 'transmissive' mode, but a radical reconfiguration took place when examinations were impending.

With regard to the final question set by Juarrero (1999), it seems very plausible that state of the system after the phase shift relates to the linguistic and pedagogical ideology associated with the certification boards. As I argued in Section 5.2.2, the certification examinations were underpinned by communicative conceptualisations of language, and the examination format, which involved the direct testing of communicative skills, prioritised communicative pedagogy. Although the reconfiguration of the system was associated with non-local linguistic and pedagogical beliefs, it is important to note that it was also beneficial to the language school itself. Through this reconfiguration, the language school provided instruction that increased the students' success rates in the examinations, which increased its own commercial viability.

In this section, I looked into the phase shift that occurs in the pedagogical activity of the language school, and used it as a springboard to develop a tentative hypothesis regarding the interplay between proximal and distal influences in ELT. This account, which draws on complexity theory, offers intriguing insights regarding the resilience of local pedagogy, and identifies one of the mechanisms through which globalising forces associated with ELT permeate the local context. Further research, which might confirm this tentative hypothesis, presents itself as a worthwhile possibility for future empirical work.

So far, I have described how an ELT setting might be conceptualised as a complex system, and discussed why such a conceptualisation may be epistemologically useful. What remains to be done is to place this study within a broader research agenda, and trace possible avenues by future research.

7.5 Future directions

A finding that recurred in this study was the tension between transmissive and communicative teaching, or (more broadly) between the technical and mainstream paradigms which were defined in Chapter 3. This tension seems to relate to the differences between the beliefs and practices associated with Greek education, and the linguistic and pedagogical ideology associated with ELT as a global profession. It is, however, unclear whether such tensions are idiosyncratic to the language school or if they recur in other ELT settings in Greece. The generalisability of these findings could perhaps be confirmed through large- scale survey research which could focus on the perceptions, beliefs and practices of teaching practitioners in Greece. Alternatively, a series of case studies in language education settings selected through theoretical criteria could provide partial confirmation, as well as theoretical depth, to the findings. Some possibilities for such research include case studies in schools in different locations (e.g., rural or metropolitan settings), schools of different sizes and with different student makeup, and ELT situations in the state education system.

Another intriguing possibility for further research involves tracing the impact of socioeconomic changes on the intentionalities that appear to underlie ELT in Greece. Fieldwork for this study was conducted in late 2010 and early 2011, which means that societal beliefs and practices that were used to derive intentionalities may no longer be current. One particular concern is that the findings reported in this thesis may not fully account for the societal impact of prolonged economic depression, the effects of which began to be felt after the austerity drives that were initiated in May 2010. A follow-up study in the language school or a similar setting might be able to provide traces of how dynamics of intentions prevailing in the Greek ELT context adjust to such severe perturbations.

The absence of information regarding the critical paradigm in the data presents two possibilities for further research. More obviously, it may prove useful to conduct research in educational settings where different paradigms co-exist. For instance, it may prove interesting to compare the tensions that arose in this study against the tensions that might emerge in settings where the critical and mainstream paradigms are more salient, as is the case in some ESL contexts. A series of ethnographic case studies, such as the one reported in this book, could be used to document different kinds of tensions, which might subsequently be integrated in a broader theory about ELT.

Apart from ethnographic investigations, it may prove interesting to conduct action research inquiries (Silver, 2008; Wallace, 1998) in educational settings similar to the language school, in order to introduce aspects of critical pedagogy into their activity. Such an approach is in line with the recommendations in the CST literature, where it is suggested that action research and experiments that deliberately perturb a system can offer insights into its potential and the underlying processes that sustain its activity (Larsen-Freeman & Cameron, 2008). Examples of such research might include projects that attempt to challenge perceptions about the normative role of the standard language, or to problematise the societal effects of ELT. Despite the fact that Action Research is praxis-oriented, the observational data generated in these projects could advance our understanding of both ELT and complex systems in general.

In this book, I argued for the feasibility and potential of a complexity-informed study of an educational setting. Using complexity as an ontological frame, this study generated a rich description of a language school in the periphery of the English-speaking world, which has extended, and at times challenged, prevailing beliefs about ELT in Greece. Using the data from the language school, it has yielded insights that advance our understanding of complex systems. While this study has made a step towards nativising complexity to the domain of applied linguistics, it is clear that considerable work remains to be done in this direction. When I presented this study as a PhD thesis, I concluded with the phrase that 'this exploration is only beginning'. In the years that have passed since then, the literature has been enriched with a small but growing number of contributions that have increasingly used complexity as a frame. Maybe then, the exploration has begun, but the feeling of excitement associated with new beginnings has not subsided.

Appendix
Methodological remarks

The purpose of this appendix is to briefly outline the empirical work that was involved in the study of the language school (see also Kostoulas, 2015a).

In brief, this study drew on the traditions of ethnography and grounded theory. Ethnographic methods were mainly used for data generation, as recommended by Richards (1996). In doing so, I aligned myself with recommendations in the literature pertaining to the study of complex social phenomena (Larsen-Freeman & Cameron, 2008), and of ELT settings (Canagarajah, 1999; Holliday, 1994). This involved the use of observational and interactional methods over an extended period of time (Fielding, 2008). However, while ethnography informed aspects of data generation, it was not used as a template for all aspects of the research design and representation.

The second methodological influence on the design of this study was grounded theory (Corbin & Strauss, 2008; Dey, 1999; Glaser & Strauss, 1967). Consistent with the principle of theoretical sampling, the empirical focus of each research phase was determined by tentative findings that had been previously generated. Similarly, the analytical categories and conceptual tools used in this study were derived through structured engagement with the data, prior to being connected to existing theoretical accounts. While broadly aligned to the grounded theory tradition, this project deviated from the recommendations of 'purist' versions of the theory (Glaser, 1992). Acknowledging that the interpretative account of the data could never be completely free of my pre-conceptions (Silverman, 2005), I capitalised on these through explicit reflexive engagement.

Obtaining access

After obtaining initial agreement with the management of the language school to conduct the study, a series of meetings were held with the management and key personnel for the negotiated delimitation of which aspects of the school life would come under study. The first meeting served primarily to introduce school stakeholders to my research. During the meeting, the school stakeholders were presented with a draft research plan and sample research instruments (questionnaires and interview guides), and we had the opportunity to discuss their concerns. Suggestions were also invited regarding the most appropriate timeframe for the research, and which aspects of the day-to-day operations of the institute they would be comfortable to have observed. A revised research plan, which incorporated the feedback from the school stakeholders, was presented in a second meeting, during which consent to conduct research was formally provided by the school management. A third meeting was held with institute staff, in order to present them with information regarding the study and invite their comments.

Generating data

In my initial research plan, provision had been made for a ten-month period of fieldwork, starting in September 2010. It was envisaged that the data generation period would be divided into four phases, which corresponded to the four nine-week terms in which an academic year was divided at the language school (Table A.1). Each phase consisted of a seven-week period of fieldwork, and two weeks of initial analysis. The direction and empirical focus of each phase was suggested by the findings that emerged in the previous one. The final phase was designed for validation, but was not implemented due to participant fatigue.

Table A.1: Data generation phases

Phase	Duration	Focus
One	September 2010 – October 2010	Content, methods, goals
Two	November 2010 – January 2011	Transmissive paradigm
Three	February 2011 – April 2011	Intentionalities, context
*Four**	*May 2011 – June 2011*	*Validation*

*not implemented

Four research strands were implemented in parallel throughout the three fieldwork phases (Table A.2). For triangulation, each research strand focused on a different source using a different method (Brown, 2001). In Strand A, eleven semi-structured interviews were used to elicit data from teachers. These data were recorded in interview transcripts and summaries, as well as methodological and reflexive memos. In Strand B, four questionnaire surveys were administered to students, from whom 60 completed questionnaires were collected. In Strand C, twelve lessons were observed, and pre- and post-observation interviews were conducted with the participating teachers. Observation notes were used to generate narrative reconstructions of the lessons. In Strand D, qualitative and quantitative methods were combined to study documentary evidence (e.g., syllabus documents, reports, learning materials etc.); the most important of these involved the content analysis of learning materials, which is mainly reported in Chapter 4. These primary sources of data were complemented by secondary data, including substantive, reflexive and methodological memos, as well as interim reports.

Table A.2: Overview of primary data

Strand	Source	Method	Primary Data
A	Teachers	Semi-structured interviews	11 transcripts/summaries 11 interview forms
B	Learners	Questionnaires	60 questionnaires
C	Lessons	Classroom Observations	12 observation forms 12 reconstructed lessons
D	Learning materials	Content Analysis	1.023 textbook activities Non-structured corpus of learning and testing materials and syllabus documents

Engaging with the data

Data analysis was carried out in four main stages: (a) preliminary analytical work; (b) open coding; (c) axial coding; and (d) theory development.

Figure A.1: Overview of preliminary analytical work

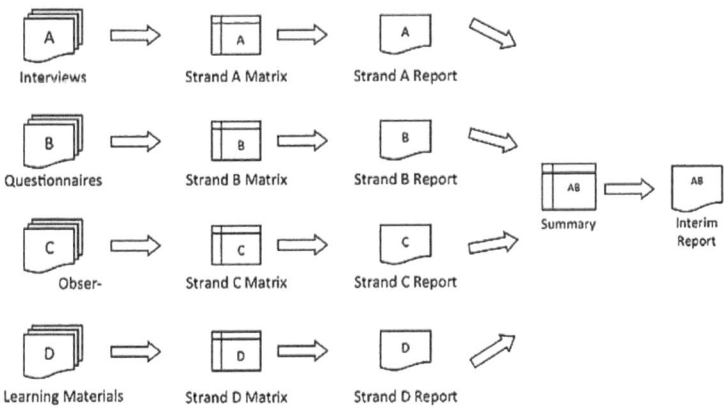

Preliminary analytical work was carried out alongside data generation between October 2010 and May 2011. The main analytical aim at this stage of analysis was to consolidate the data, and to synthesise the emergent findings from the four different strands into a coherent description of the language school. This was achieved through alternating stages of data display and analytical text generation (Miles & Huberman, 1994), as shown in Figure A.1. At the end of every nine-week data generation phase, the information from each data source (i.e., teachers, students, lessons, documents) was entered into a matrix to facilitate comparison and contrast of emerging themes. Each matrix served as the basis for the generation of a report that summarised the data from that particular strand of the study. After the four strand-specific reports had been completed, the process was iterated in order to synthesise the emergent findings. A summary matrix was created, where findings from the different strands were displayed, and an interim progress report was then derived from it.

After the data had been generated, **open coding** was used to generate an initial set of categories from my dataset, which now included primary sources (Table A.2) and secondary sources (interim reports, analytical and reflexive memos etc.). In practical terms, this involved reading through the data and assigning one or more conceptual labels to various thematic units with the help of Computer Assisted Qualitative Data Analysis Software. A rough coding scheme had already been derived during preliminary analysis, and to ensure its consistent application, I gradually produced a codebook in which each category was described. The scheme was constantly refined, as new categories were identified, and others were conflated, dropped or expanded with new subcategories. After several iterations, the coding scheme was stabilised, and short memos were produced summarising the information in each category. These memos served as a permanent record of the process that led to the definition of the category, as well as a stimulus for further reflection. Additional memos were generated, where methodological decisions and reflexive comments were recorded.

The third phase of analysis corresponded to the grounded theory process of 'theoretical' (Dörnyei, 2007) or 'axial' coding (Corbin & Strauss, 2008), during which connections were drawn between the previously defined categories. Initially, fourteen major categories, or axes, were abstracted from the data (Table A.3, left). In order to facilitate conceptual manipulation, these axes were grouped into three super-ordinate categories. Using a variety of analytical tools and processes, hierarchical, associative and causal connections were drawn between these categories. These connections and other insights that emerged from analytical engagement with the data were recorded in eight extended analytical memos. Each memo, which ranged in length from 3,500 to 13,000 words, contained information on an axis or a super-ordinate category, resulting in a 55,000-word long corpus of analytical text.

Subsequent engagement with the analytical text resulted in further improvement of the coding scheme, and part of the dataset was re-coded using a refined set of categories (Table A.3, middle). The revised categories were used to provide organisational structure to the study: categories pertaining to the overarching construct of coursebook content are discussed in Chapter 4, categories pertaining to intentionalities are presented in Chapter 5, and categories pertaining to teaching and learning activities are described in Chapter 6 (Table A.3, right).

Table A.3: Coding schemes used for axial coding

Initial coding scheme	Final coding scheme	Chapter
Content of instruction (What?) • Pronunciation • Grammar • Vocabulary • Speaking • Listening • Reading • Writing	Coursebook content • Grammar • Vocabulary • Oral skills • Reading • Writing	4
Ends of instruction (Why?) • Certification • Integration • Culture • Supplementation • Protectionism	Intentionalities • Certification • Integration • Culture • Competition • Protectionism	5
Methods of instruction (How?) • Examinations • Coursebooks • Language of instruction	Learning activities • Reading and Vocabulary • Traditional Grammar • Inductive Grammar • Genre-based writing • Process-based writing • Integrated oral skills • Review	6

Theory generation, which concluded the data analysis, consisted of two processes: conceptualisation and theoretical validation. Conceptualisation referred to the process of developing new concepts, or refining existing ones, in order to better account for the data (Punch, 2005; Silverman, 2005). Theoretical validation involved using existing theory, predominantly from CST, as a conceptual touchstone against which my emergent understandings were compared. To illustrate, the inductively derived construct of 'ends, aims or purposes of instruction' was related to the construct of intentionality from the literature (Stelma, 2011; Young et al. 2002), whereas the 'regularly occurring but unpredictable patterns of behaviour' were related to the construct of attractors from CST (Larsen-Freeman & Cameron, 2008). The end product of these processes was a number of theoretical constructs that extend our conceptual toolkit for understanding ELT. These constructs, namely intentionality, the affordance landscape, and prototypical lessons, are described in more detail in Chapters 4, 5 and 6 respectively.

Presentation of data

One particularly challenging decision that had to be made in preparing this book was deciding how to deal with bilingual data. On the one hand, the data had to be made accessible to readers who might not understand Modern Greek, and on the other I was keen to remain faithful to the participants' voices and preserving interpretative validity (Maxwell, 2002). The tension between these two competing aims, which was further compounded by the ethical and political implications of representation, was resolved through a representational strategy that was as flexible as possible.

With English language data, my representation choices were often dictated by the principle of non-malfeasance. Although it would sometimes seem desirable to reproduce the participants' output with maximal faithfulness, such a representational strategy risked stigmatising the participants. For instance, some participants who were non-native speakers of English occasionally used non-standard language forms such as: 'For example (2 sec) I when I <u>teached</u> the passive voice, I gave them the rule, and I told them "try to give me an example"' (emphasis added). Given the high premium placed on linguistic accuracy in the professional context of the research participants, the publication of any instances of non-standard use was expected to have detrimental effects on the participants' confidence and professional standing. It was also likely to negatively impact the reputation of the school where the research took place. In such cases, after analysis but prior to representation, the data were subjected to unobtrusive editing, at which stage language was standardised. This was not done with the language learners' data, however, as the risk of stigmatisation was lower (with the possible exception of their mothers, nobody really expected perfection from the students), and did not justify sacrificing authenticity. In these cases, I have preserved the learners' original wording and spelling, and indicated deviations from the standard with the notation '[sic]'.

With regard to Greek language data, I opted for presenting a translation in English alongside the original. This choice was partly motivated by a desire to promote the visibility of languages other than English in academic discourse and countering the hegemonic status of English as an academic lingua franca. An additional motive was the concern that translations tended to conceal linguistic behaviour that could be theoretically significant. The following extract, from an interview with one of the teachers, illustrates this point:

Teacher:	Και το ξέρεις κι εσύ καλύτερα από μένα, πως στην σχ- >το μάθαμε και απ' την σχολή< ότι για να μάθει το παιδί κάτι πρέπει απαραίτητα να γίνει κάποιο «<u>noticing</u>», έτσι; (2 sec) Αλλά ο Άγγλος όταν μιλάει με σωστή, με <u>RP</u>, θα πει «<u>dishes</u>» αλλά επειδή είναι <u>native speaker</u>, θα είναι με <u>devoicing</u> στο d, ξέρεις, και καλά σαν <u>tishes</u> (3 sec)
Achilleas:	Πιστ- Νομίζεις ότι αυτό είναι, ότι αποτελεί πρόβλημα;
Teacher:	Ναι, αλλά ο Άγγλος αυτό το κάνει χωρίς να το καταλαβαίνει, γιατί είναι <u>native speaker</u>. Και όχι μόνο αυτό, δεν καταλαβαίνει κιόλας ότι αυτό είναι, συνιστά πρόβλημα για τον Έλληνα μαθητή, αν μιλάει έτσι.

An unreflective translation of such a transcript would obscure the high prevalence of English lexical items (bolded). Presenting such a translation without reference to the original might divert attention from the functional distribution of languages (i.e., under what circumstances each language tended to be used) at the language school, and from the way in which this teacher constructed her professional identity by using English language terms when discussing teaching methodology. Other theoretically significant issues, which translations risked masking, included dialectic variation (in Greek and in English), and the effects of code switching. Though such issues lie outside the purview of this study, it was felt that, in the interest of "validity through transparency and access" (Nikander, 2008, p. 227), readers should have access to the original text as well as a translation whenever possible.

References

Abbot, Garry. "Encouraging communication in English: A paradox". *ELT Journal,* 35(3), 1981, pp. 228–230.

Adamo, Grace Ebunlola. "Nigerian English". *English Today,* 23(1), 2007, pp. 42–47.

Agar, Micheal. "We have met the other and we're all non-linear: Ethnography as a non-linear dynamic system". *Complexity,* 10(2), 2004, pp. 16–24.

Agathopoulou, Eleni. "EFL student teachers' beliefs and the effect of a second language acquisition course". Paper presented at the 14th International Conference of the Greek Association of Applied Linguistics. Thessaloniki, Greece. 2007. Retrieved on 9 August 2014 from: http://www.enl.auth.gr/gala/14th/Papers/English%20papers/Agathopoulou.pdf.

Alderson, J. Charles / Clapham, Caroline / Wall, Dianne. *Language test construction and evaluation.* Cambridge: Cambridge University Press. 1995.

Alexandropoulou, M. *Teaching grammar to EFL learners in Greek state schools.* Unpublished M.Ed. dissertation, Hellenic Open University, Patras, Greece. 2002.

Alexiou, Thomais / Mattheoudakis, Marina. "Introducing a foreign language at primary level: Benefits or lost opportunities? The case of Greece". *Research Papers in Language Teaching & Learning,* 4(1), 2013, pp. 99–119.

Aliakbari, Mohammad. "The place of culture in the Iranian ELT textbooks". Paper presented at the *9th Conference of Pan-Pacific Association of Applied Linguistics.* 2004. Retrieved 9.8.2014 from: http://www.paaljapan.org/resources/proceedings/PAAL9/pdf/Aliakbari.pdf.

Allwright, Dick. "Exploratory practice: Rethinking practitioner research in language teaching". *Language Teaching Research,* 7(2), 2003, pp. 113–141.

Allwright, Dick. "Developing principles for practitioner research: The case of exploratory practice". *The Modern Language Journal,* 89(3), 2005, pp. 353–366.

Allwright, Dick / Hanks, Judith. *The developing language teacher: An introduction to exploratory practice.* London: Palgrave Macmillan. 2009.

Alptekin, Cem. "Target-language culture in EFL materials". *ELT Journal,* 47(2), 1993, pp. 136–143.

Alptekin, Cem. "Towards intercultural communicative competence in ELT". *ELT Journal,* 56(1), 2002, pp. 57–64.

Anderson, Anne / Lynch, Tony. *Listening.* Oxford: Oxford University Press. 1998.

Anderson, Ruth A. / Crabtree, Benjamin F. / Steele, David J. / McDaniel, Reuben R. "Case study research: The view from complexity science". *Qualitative Health Research*, 15(5), 2005, pp. 669–685.

Angouri, Jo / Mattheoudakis, Marina / Zigrika, Maria. "Then how will they get the 'much-wanted paper'?: A multifaceted study of English as a foreign language in Greece." *Advances in research on language acquisition and teaching: Selected papers (Proceedings of the 14*th *International Conference of Greek Applied Linguistics Association)*. Athens: Greek Association of Applied Linguistics. 2010, pp. 179–194.

Argyropoulos, Vasileios M. *Φωνηεντιάδα: Μια δεύτερη ματιά. [The vowel saga: A second look]*. Athens: Carpe Librum. 2016.

Aslan, Erhan. "When the native is also a non-native: 'Retrodicting' the complexity of language teacher cognition". *Canadian Modern Language Review*, 71(3), 2015, pp. 244–269.

Austin, John Langshaw. *How to do things with words. The William James lectures delivered at Harvard University in 1955*. London: Clarendon Press. 1962.

Baba, Kyoko / Nitta, Ryo. "Phase transitions in development of writing fluency from a complex dynamic systems perspective". *Language Learning*, 64(1), 2014, pp. 1–35.

Barab, Sasha A. / Cherkes-Julkowski, Miriam / Swenson, Rod / Garrett, Steve / Shaw, Robert E. / Young, Michael. "Principles of self-organization: Learning as participation in autocatakinetic systems". *Journal of the Learning Sciences*, 8(3–4), 1999, pp. 349–390.

Bax, Stephen. "Roles for a teacher educator in context-sensitive teacher education". *ELT Journal*, 51(3), 1997, pp. 232–241.

Bax, Stephen. "The end of CLT: A context approach to language teaching". *ELT Journal*, 57(3), 2003, pp. 278–287.

Baxter, G. J. / Blythe, R. A. / Croft, W. / McKane, A. J. "Utterance selection model of language change". *Physical Review E*, 73(4), 2006, article 046118.

Beckner, Clay / Blythe, Richard / Bybee, Joan / Christiansen, Morten H. / Croft, William / Ellis, Nick C. / Holland, John / Ke Jinyun / Larsen-Freeman, Diane / Schoenemann, Tom [The Five Graces Group]. "Language is a complex adaptive system: Position paper". *Language Learning*, 59, 2009, pp. 1–26.

Bennett, Andrew / Elman, Colin. "Complex causal relations and case study methods: The example of path dependence". *Political Analysis*, 14(3), 2006, pp. 250–267.

Berns, Margie / Barret, Jeanelle / Chan, Chak / Chikuma, Yoshiki / Friedrich, Patricia / Hadjidimos, Olga-Maria / Harney, Jill / Hislope Kristi / Johnson David / Kimball, Suzanne / Low, Yvonne / McHenry Tracey / Palaiologos

Vivienne / Petray Marnie / Shapiro Rebecca / Shook, Ana Ramirez. "(Re)experiencing hegemony: The linguistic imperialism of Robert Phillipson". *International Journal of Applied Linguistics,* 9(1), 1998, pp. 138–141.

Bezantakos, N. / Papathomas, A. / Loutrianaki, E. / Charalambakos, V. *Αρχαία Ελληνική Γλώσσα Α' Γυμνασίου (Βιβλίο Εκπαιδευτικού) [Ancient Greek lanuguage, 1st Form, teacher's book].* Athens: OEDB. 2008.

Biesta, Gert. "Five theses on complexity reduction and its politics". In: Osberg, Deborah / Biesta, Gert (Eds.), *Complexity theory and the politics of education.* Rotterdam: Sense Publishers. 2010, pp. 5–14.

Biesta, Gert / Osberg, Deborah. "Complexity, education and politics from the inside-out and the outside-in". Osberg, Deborah / Biesta, Gert (Eds.), *Complexity theory and the politics of education.* Rotterdam: Sense Publishers. 2010, pp. 1–4

Bisong, Joseph. "Language choice and cultural imperialism: A Nigerian perspective". *ELT Journal,* 49(2), 1995, pp. 122–132.

Bogg, Jan / Geyer, Robert (Eds.). *Complexity, science and society.* Oxford: Raddcliffe Publishing. 2007.

Bourdieu, Pierre. *Practical reason: On the theory of action.* Stanford, CA.: Stanford University Press. 1998.

Bronfenbrenner, Uri. *The ecology of human development: Experiments by nature and design.* Cambridge, MA: Harvard University Press. 1979.

Bronfenbrenner, Uri. "Ecological systems theory". *Annuals of Child Development,* 6, 1989, pp. 187–251.

Bruthiaux, Paul. "Squaring the circles: Issues in modeling English worldwide". *International Journal of Applied Linguistics,* 13(2), 2003, pp. 159–178.

Burns, Anne / Knox, J. "Classrooms as complex systems: An emergent research agenda". Paper presented at AILA World Congress, Madison, WI. 2005.

Byram, Michael / Esarte-Sarries, Veronica / Taylor, Susan. *Cultural studies and language learning: A research report.* Clevedon: Multilingual Matters. 1991.

Byram, Michael / Feng, Anwei. "Culture and language learning: Teaching, research and scholarship". *Language Teaching,* 37(3), 2004, pp. 149–168.

Byrne, David. *Complexity theory and the social sciences: An introduction.* London: Routledge. 1998.

Byrne, David. *Interpreting quantitative data.* London: SAGE. 2002.

Byrne, David. "Complexity, configurations and cases". *Theory, Culture & Society,* 22(5), 2005, pp. 95–111.

Byrne, David. "Complex realist and configurational approaches to cases: A radical synthesis". In: Byrne, David / Ragin, Charles (Eds.), *SAGE handbook of case based methods*. London: SAGE. 2009.

Byrne, David / Callaghan, Gillian. *Complexity theory and the social sciences: The state of the art*. New York: Routledge. 2014.

Cameron, Lynne. *Teaching languages to young learners*. Cambridge: Cambridge University Press. 2001.

Cameron, Lynne / Deignan, Alice. "The emergence of metaphor in discourse". *Applied Linguistics*, 27(4), 2006, pp. 671–690.

Cameron, Lynne J. / Stelma, Juurd H. "Metaphor clusters in discourse". *Journal of Applied Linguistics*, 1(2), 2004, pp. 107–136.

Canagarajah, Suresh A. *Resisting linguistic imperialism in English teaching*. Oxford: Oxford University Press. 1999.

Canagarajah, Suresh A. "Globalisation, method and practice in periphery classrooms". In: Block, David / Cameron, Deborah (Eds.), *Globalisation and language teaching*. London: Routledge. 2002, pp. 134–150.

Canagarajah, Suresh A. *Reclaiming the local in language policy and practice*. London: Lawrence Erlbaum Associates. 2005.

Canagarajah, Suresh A. "TESOL at forty: What are the issues?" *TESOL Quarterly*, 40(1), 2006, pp. 9–34.

Canale, Michael / Swain, Merril. "Theoretical bases of communicative approaches to second language teaching and testing". *Applied Linguistics*, I(1), 1980, pp. 1–47.

Chatzidemou, D. Προετοιμασία και σχέδιο μαθήματος *[Preparation and lesson planning]*. Thessaloniki: Kyriakides. 1988.

Chan, Letty / Dörnyei, Zoltán / Henry, Alastair. "Learner archetypes and signature dynamics in the language classroom: A retrodictive qualitative modelling approach to studying L2 motivation". In: Dörnyei, Zoltán / MacIntyre, Peter D. / Henry, Alastair (Eds.), *Motivational dynamics in language learning*. Bristol: Multilingual Matters. 2014, pp. 238–259.

Chomsky, Noam. "A review of B. F. Skinner's Verbal Behavior". *Language*, 35(14), 1959, pp. 26–58.

Chomsky, Noam. *Aspects of the theory of syntax*. Cambridge, MA: M.I.T. Press. 1965.

Christias, I. Θεωρία και μεθοδολογία της διδασκαλίας *[Theory and methodology of teaching]*. Athens: Gregoris. 1992.

Cilliers, Paul. *Complexity and postmodernism*. London: Routledge. 1998.

Cilliers, Paul. "Boundaries, hierarchies and networks in complex systems". *International Journal of Information Management,* 5(2), 2001, pp. 135–147.

Cogo, Alicia. "English as a Lingua Franca: Concepts, use, and implications". *ELT Journal,* 66(1), 2012, pp. 97–105.

Corbin, Juliet M. / Strauss, Anselm L. *Basics of qualitative research: Techniques and procedures for developing grounded theory.* Thousand Oaks, CA: SAGE. 2008³.

Cortazzi, Martin / Shen, Wei-Wei. "Cross-linguistic awareness of cultural keywords: A study of Chinese and English speakers". *Language Awareness,* 10(2-3), 2001, pp. 125–142.

Council of Europe. *Common European Framework of Reference for Languages: Learning, teaching, assessment.* Cambridge: Cambridge University Press. 2001.

Crystal, David. "On trying to be Crystal-clear: A response to Phillipson". *Applied Linguistics,* 21(3), 2000, pp. 415–423.

Crystal, David. *English as a global language.* Cambridge: Cambridge University Press. 2003².

Crystal, David. "Two thousand million?" *English Today,* 24(1), 2008, pp. 3–6.

David, Maya / Govindasamy, Subra. "Negotiating a language policy for Malaysia: Local demand for affirmative action versus challenges from globalisation". In: Canagarajah, Suresh A. (Ed.), *Reclaiming the local in language policy and practice.* Mahwaj, NJ: Lawrence Erlbaum Associates. 2005, pp. 123–146.

Davies, Alan. "Ironising the myth of linguicism". *Journal of Multilingual & Multicultural Development,* 17(6), 1996, pp. 485–496.

Davies, Alan. "Standard English: Discordant voices". *World Englishes,* 18(2), 1999, pp. 171–186.

Davis, Brent / Sumara, Dennis J. *Complexity and education: Inquiries into learning, teaching, and research.* London: Lawrence Erlbaum Associates. 2006.

Davis, Kathryn A. / Lazaraton, Anne (Eds.) Qualitative research in ESOL [Special Issue]. *TESOL Quarterly,* 29(3), pp. 423–626.

de Bot, Kees / Lowie, Wander / Thorne, Steven L. / Verspoor, Marjolijn. "Dynamic systems theory as a comprehensive theory of second language development". In: Garcia Mayo, Maria del Pilar / Gutierrez Mangado, Maria Junkal / Martinez Adrián, Maria (Eds.), *Contemporary Approaches to Second Language Acquisition.* Amsterdam: John Benjamins. 2013, pp,. 199–220.

de Bot, Kees / Lowie, Wander / Verspoor, Marjolijn. "A dynamic systems theory approach to second language acquisition". *Bilingualism: Language and cognition,* 10(1), 2007, pp. 7–21.

DeLanda, Manuel. *Intensive science and virtual philosophy*. London: Continuum. 2005

Dendrinos, Bessie. "Comment 3 to 'Lingua franca or lingua frankensteinia'". In: Phillipson, Robert (Ed.), *Linguistic imperialism continued*. London: Routledge. 2009, pp. 181–182.

Descartes, René. *Discours de la méthode pour bien conduire sa raison, et chercher la vérité dans les sciences (avec introduction et notes par Êtienne Gilson) [A discourse on method for conducting one's reason well and searching for truth in the sciences (intr. & notes by Êtienne Gilson]*. Paris: Librairie Philosophique J. Vrin. 1637 / 1966

Dewey, Martin. "Towards a post-normative approach: Learning the pedagogy of ELF". *Journal of English as a Lingua Franca,* 1(1), 2012, pp. 141–170.

Dewey, Martin. "The distinctiveness of English as a Lingua Franca". *ELT Journal,* 67(3), 2013, pp. 346–349.

Dey, Ian. *Grounding grounded theory: Guidelines for qualitative inquiry*. San Diego, CA: Academic Press. 1999.

Dong, Jihua. "A dynamic systems theory approach to development of listening strategy use and listening performance". *System,* 63, 2016, pp. 149–165.

Dörnyei, Zoltán. *The psychology of the language learner: Individual differences in second language acquisition*. Mahwah, NJ: Lawrence Erlbaum. 2005.

Dörnyei, Zoltán. *Research methods in applied linguistics: Quantitative, qualitative, and mixed methodologies*. Oxford: Oxford University Press. 2007.

Dörnyei, Zoltán. "The English language and the Word of God". In: Wong, Mary Shepard / Canagarajah, Suresh, A. (Eds.), *Christian and critical English language educators in dialogue: Pedagogical and ethical dilemmas*. New York: Routledge. 2009, pp. 154–157.

Dörnyei, Zoltán. "Researching complex dynamic systems: 'Retrodictive qualitative modelling' in the language classroom". *Language Teaching,* 47(1), 2014, pp. 80–91.

Dörnyei, Zoltán / Csizér, Kata. "Ten commandments for motivating language learners: Results of an empirical study" *Language Teaching Research,* 2(3), 1998, pp. 203–229.

Edge, Julian. "Imperial troopers and servants of the Lord: A vision of TESOL for the 21st century". *TESOL Quarterly,* 37(4), 2003, pp. 701–709.

Edwards, Alison. "Dutch English: Tolerable, taboo, or about time too?" *English Today,* 26(1), 2010, pp. 19–25.

Ellis, Rod. *Task-based language learning and teaching*. Oxford: Oxford University Press. 2003.

Engels, F. *Dialektik der Natur [Dialectics of Nature]*. Berlin: Holzinger. 1883 / 2011.

Eve, Raymond A. / Horsfall Sara / Lee, Mary E. (Eds.). *Chaos, Complexity, and Sociology: Myths, Models, and Theories*. Thousand Oaks, CA: SAGE. 1997.

Exarchopoulos, N. I. *Ειδική διδακτική [Particular didactics]*. Athens: Kontos-Fylaktos. 1962.

Fairclough, Norman. *Critical discourse analysis: The critical study of language*. London: Longman. 1995.

Fay, Richard / Lytra, Vally / Ntavaliagkou, Maria. "Multicultural awareness through English: A potential contribution of TESOL in Greek school". *Intercultural Education*, 21(6), 2010, pp. 579–593.

Feryok, Anne. "Language teacher cognition: An emergent phenomenon in an emergent field". In: Mercer, Sarah / Kostoulas, Achilleas (Eds.), *Language Teacher Psychology*. Bristol: Multilingual Matters. 2018, pp. 105–121.

Fielding, Nigel. "Ethnography". In: Gilbert, Nigel, G. (Ed.), *Researching social life*. Thousand Oaks, CA: SAGE. 2008³, pp. 266–285.

Firth, Alan. "The discursive accomplishment of normality: On 'lingua franca' English and conversation analysis". *Journal of Pragmatics*, 26(2), 1996, pp. 237–259.

Fishman, J. A. "Review of Linguistic Imperialism". *The Modern Language Journal*, 77(3), 1993, pp. 399–400.

Gardner, Robert C. / Lambert, Wallace E. *Attitudes and motivation in second language learning*. Rowley, MA: Newbury House. 1972.

Geeraerts, Dirk. "Prospets and problems of prototype theory". In: Geeraerts, Dirk (Ed.), *Cognitive linguistics: Basic readings*. The Hague: Mouton de Gruyter. 2006.

Georgiadi, E. *The importance of understanding ELT methodology: Design and procedure*. Unpublished M.Ed. dissertation. Hellenic Open University. Patras, Greece. 2003.

Ghosn, Irma K. "Four good reasons to use literature in primary school ELT". *ELT Journal*, 56(2), 2002, pp. 172–179.

Giannoulis, N. I. *Εισαγωγή στη γενική διδακτική [Introduction to general didactics]* (2nd ed.). Athens: publisher not indicated. 1980.

Gibson, James J. "The theory of affordances". In: Shaw, Robert / Bransford, John (Eds.), *Perceiving, acting and knowing: Towards an ecological psychology*. Hillsdale, NJ.: Lawrence Erlbaum. 1977, pp. 67–82.

Gibson, James J. *The ecological approach to visual perception*. Boston, MA: Houghton Mifflin. 1979.

Glaser, Barney G. *Basics of grounded theory analysis: Emergence vs forcing*. Mill Valley, CA: Sociology Press. 1992.

Glaser, Barney G. / Strauss, Anselm L. *Discovery of grounded theory: Strategies for qualitative research*. New Brunswick, NJ: AldineTransaction. 1967.

Gleick, James. *Chaos: Making a new science*. New York: Penguin Books. 1987.

Gomm, Roger. *Normal results: An ethnograpic study of health visitor student assessment*. Unpublished PhD thesis, The Open University, Milton Keynes. 1986.

Gomm, Roger. "Uncertain minds or uncertain times?". In: Gomm, Roger / Needham, Gill / Bullman, Anne (Eds.), *Evaluating research in health and social care*. Thousand Oaks, CA: SAGE. 2000, pp. 230–238.

Gomm, Roger / Hammersley, Martyn. "Thick ethnographic description and thin models of complexity". Paper presented at the Annual Conference of the British Educational Research Association, Leeds. 2001.

Graddol, David. *The future of English?: A guide to forecasting the popularity of the English language in the 21st century*. London: British Council. 1997.

Graddol, David. *English next*. London: British Council. 2006.

Griffin, Jeffrey L. "Global English infiltrates Bulgaria". *English Today*, 17(4), 2001, pp. 54–60.

GSEE Centre for the Development of Educational Policy. *Δημόσιες και ιδιωτικές δαπάνες για την εκπαίδευση σε περιβάλλον κρίσης [Public and private educational expenditure at a time of crisis]*. Athens: Author. 2014.

Guariento, William / Morley, John. "Text and task authenticity in the EFL classroom". *ELT Journal*, 55(4), 2001, pp. 347–353.

Guba, Egon G. / Lincoln, Yvonna S. "Competing paradigms in qualitative research". In: Denzin, Norman K. / Lincoln, Yvonna S. (Eds.), *Handbook of qualitative research*. Thousand Oaks, CA: SAGE. 1994, pp. 105–117.

Häggblom, Charlotta. *Young EFL-pupils reading multicultural children's fiction: An ethnographic case study in a Swedish language primary school in Finland*. Åbo, Finland: Åbo Akademi University Press. 2006.

Haggis, Tamsin. "Conceptualising the case in adult and higher-education research: A dynamic systems view". In: Bogg, Jan / Geyer, Robert (Eds.), *Complexity, Science & Society*. Oxford: Radcliffe. 2007, pp. 38–43.

Haggis, Tamsin. "'Knowledge must be contextual': Some possible implications of complexity and dynamical systems theories for educational research". In: Mason, Mark (Ed.), *Complexity theory and the philosophy of education*. Oxford: Wiley-Blackwell. 2008.

Halai, Nelofer. "Making use of bilingual interview data: Some experiences from the field". *The Qualitative Report*, 12(3), 2007, pp. 344–355.

Halliday, Michael Alexander Kirkwood. *Explorations in the functions of language.* London: Edward Arnold. 1973.

Hanks, William F. *Language & communicative practices.* Oxford: Westview Press. 1996.

Harmer, Jeremy. *The practice of English language teaching.* Harlow: Longman. 2015[5].

Hedström, Peter. *Dissecting the social: On the principles of analytical sociology.* Cambridge: Cambridge University Press.

Henry, Alastair. "Conceptualizing teacher identity as a complex dynamic system: The inner dynamics of transformations during a practicum". *Journal of Teacher Education,* 67(4), 2016, pp. 291–305.

Herdina, Philip / Jessner, Ulrike. *A dynamic model of multilingualism.* Clevedon: Multilingual Matters. 2002.

Hiver, Phil / Al-Hoorie, Ali H. "A dynamic ensemble for second language research: Putting complexity theory into practice". *The Modern Language Journal, 100*(4), 2016, pp. 741–756.

Hohenberger, Annette. *Functional categories in language acquisition: Self-organization of a dynamical system.* Tübingen: Niemeyer. 2002.

Hohenberger, Annette. / Peltzer-Karpf, Annemarie. "Language learning from the perspective of nonlinear dynamic systems." *Linguistics,* 47(2), 2009, pp. 481–511.

Holborow, M. "Review of Linguistic Imperialism". *ELT Journal,* 47(4), 1993, pp. 358–360.

Holliday, Adrian. "Tissue rejection and informal orders in ELT projects: Collecting the right information". *Applied Linguistics,* 13(4), 1992, pp. 403–424.

Holliday Adrian. *Appropriate methodology and social context.* Cambridge: Cambridge University Press. 1994.

Holliday, Adrian. "Developing a sociological imagination: Expanding ethnography in international English language education". *Applied Linguistics,* 17(2), 1996, pp. 234–255.

Holliday, Adrian. "Small cultures". *Applied Linguistics,* 20(2), 1999, pp. 237–264.

Holliday, Adrian. *The struggle to teach English as an international language.* Oxford: Oxford University Press. 2005.

Hopwood, Nick "Doctoral experience and learning from a sociocultural perspective". *Studies in Higher Education,* 35(7), 2010, pp. 829–843.

Howatt, A. P. R. *A history of English language teaching.* Oxford: Oxford University Press. 2004[2].

Hu, Guangwey. "Potential cultural resistance to pedagogical imports: The case of Communicative Language Teaching in China". *Language, Culture & Curriculum,* 15(2), 2002, pp. 93–105.

Hughes, Arthur / Lascaratou, Chrysoula. "Competing criteria for error gravity". *ELT Journal,* 36(3), 1982, pp. 175–182.

Hutchinson, Tom / Torres, Eunice. "The textbook as agent of change". *ELT Journal,* 48(4), 1994, pp. 315–328.

Hymes, Dell H. "On communicative competence". In: Pride, J. B. / Holmes, Janet (Eds.), *Sociolinguistics: Selected readings.* Harmondsworth: Penguin. 1972, pp. 269–293.

Irvine-Nikiaris, Christine. *ECPE: Teacher's guide.* Athens: Hellenic American Union. 2009.

Ishida, Midori. "Effects of recasts on the acquisition of the aspectual form -te i- (ru) by learners of Japanese as a foreign language". *Language Learning,* 54(2), 2004, pp. 311–394.

Jacob, G. P. *Coming to terms with imperialism: Safeguarding local knowledge and experience in ELT curriculum development.* Unpublished manuscript. Department of English, University of Pune, India. 1996

Jacobs, George M. / Ball, Jessica. "An investigation of the structure of group activities in ELT coursebooks". *ELT Journal,* 50(2), 1996, pp. 99–107.

Jenkins, Jennifer. *The phonology of English as an international language: New models, new norms, new goals.* Oxford: Oxford University Press. 2000.

Jenkins, Jennifer. "A sociolinguistically based, empirically researched pronunciation syllabus for English as an International Language". *Applied Linguistics,* 23(1), 2002, pp. 83–103.

Jenkins, Jennifer. "Current perspectives on teaching World Englishes and English as a Lingua Franca". *TESOL Quarterly,* 40(1), 2006a, 157–181.

Jenkins, Jennifer. "Global intelligibility and local diversity: possibility or paradox?" In: Rubdy, Rani / Saraceni, Mario (Eds.), *English in the world: Global rules, global roles.* London: Continuum. 2006b, pp. 32–39.

Jenkins, Jennifer. "Points of view and blind spots: ELF and SLA". *International Journal of Applied Linguistics,* 16(2), 2006c, pp. 137–162.

Jenkins, Jennifer. "The spread of EIL: A testing time for testers". *ELT Journal,* 60(1), 2006d, pp. 42–50.

Jenkins, Jennifer. *English as a lingua franca: Attitude and identity.* Oxford: Oxford University Press. 2007.

Jenkins, Jennifer / Cogo, Alicia / Dewey, Martin. "Review of developments into research into English as a lingua franca". *Language Teaching*, 44(3), 2011, pp. 281–215.

Jenks, Christopher. "Are you an ELF? The relevance of ELF as an equitable social category in online intercultural communication". *Language & Intercultural Communication*, 13(1), 2013, pp. 95–108.

Jessner, Ulrike. "A DST model of multilingualism and the role of metalinguistic awareness". *The Modern Language Journal*, 92(2), 2008, pp. 270–283.

Johnson, Neil F. *Simply complexity: A clear guide to complexity theory*. Oxford: Oneworld. 2009.

Johnson, Karen E. / Golombek, Paula, R. "The transformative power of narrative in second language teacher education." *TESOL Quarterly*, 45(3), 2011, pp. 486–509.

Juarrero, Alicia. *Dynamics in action: Intentional behaviour as a complex system*. Cambridge, MA: M.I.T. Press. 1999.

Kachru, Braj B. "Standards, codification and socio-linguistic realism: The English language in the outer circle". In: Quirk, Randolph / Widdowson, Henry G. (Eds.), *English in the world: Teaching and learning the language and literatures*. Cambridge: Cambridge University Press. 1985, pp. 11–30.

Kachru, Braj B. "Liberation linguistics and the Quirk concern". *English Today*, 7(1), 1991, pp. 3–13.

Kachru, Yamuna. "'Interlanguage and language acquisition research' Review of L. Selinker: Rediscovering Interlanguage". *World Englishes*, 12(1), 1993, pp. 265–268.

Kachru, Yamuna. "Teaching and learning World Englishes". In Hinkel, Eli. (Ed.), *Teaching and learning of World Englishes*. Mahwah, NJ: Lawrence Erlbaum Associates. 2005.

Kaloscai, K. "Erasmus exchange students: A behind the scenes view into an ELF community of practice". *Apples – Journal of Applied Language Studies*, 3(1), 2009, pp. 25–49.

Kanellou, Vasiliki. "Accents of English and listening comprehension: Evidence of conflict between ELT handbooks and teacher's practices". Paper presented at the 44th Annual British Association of Applied Linguistics meeting: 'The Impact of Applied Linguistics', Cardiff. 2012.

Karagianni, Evangelia. *Teacher development and emotions: An ICT-based reflective/collaborative approach*. Unpublished PhD thesis, University of Athens. 2012. Retrieved 20.8.2014 from http://www.didaktorika.gr/eadd/handle/10442/31746

Karavas-Doukas, Evdokia. "Teacher identified factors affecting the implementation of an EFL innovation in Greek public secondary schools". *Language, Culture and Curriculum*, 8(1), 1995, pp. 53–68.

Karavas, Evdokia. "How satisfied are Greek EFL teachers with their work? Investigating the motivation and job satisfaction levels of Greek EFL Teachers". *Porta Linguarum*, 14, 2010, pp. 59–78.

Karavas, Evdokia. "Implementing innovation in primary EFL: A case study in Greece". *ELT Journal*, 68(3), 2014, pp. 243–253.

Karavas, Kia. *The KPG speaking test in English: A handbook*. Athens: Research Centre for the English Language. 2009.

Kim, Jaegwon. "Making sense of emergence". *Philosophical studies*, 95(1–2), 1999, pp. 3–36.

King, Jim. *Silence in the second language classroom*. Basingstoke: Palgrave-MacMillan. 2013.

Kingman, John. Report of the committee of inquiry into the teaching of English language. London: Her Majesty's Stationary Office. 1988.

Kirkpatrick, Andy. *English as a lingua franca in ASEAN: A multilingual model*. Hong Kong: Hong Kong University Press. 2010.

Kirkpatrick, Andy. "English as an Asian lingua franca and the multilingual model of ELT". *Language Teaching*, 44(2), 2011, pp. 212–224.

Kitsios, K. I. *Ειδική διδακτική [Particular didactics]*. Ioannina: Dodone. 1992.

Klimpfinger, Theresa V. "'Mind you, sometimes you have to mix' – The role of codeswitching in English as a lingua franca". *Vienna English Working Papers*, 16(2), 2007, 36–61.

Kortmann, Berndt / Szmrecsanyi, Benedict. *Linguistic complexity: Second language acquisition, indigenization, contact*. Berlin: Walter de Gruyter. 2012.

Kostoulas, Achilleas. *Dynamics impacting ELT learning materials in the Greek context: A systems perspective*. Unpublished M.A. dissertation, The University of Manchester, UK. 2007.

Kostoulas, Achilleas. «Τα Αγγλικά υπό το πρίσμα της πολυπολιτισμικότητας: Κριτική επισκόπηση των εγχειριδίων Αγγλικής γλώσσας στο Γυμνάσιο και εναλλακτικές προτάσεις». [English under the multicultural lens: A critical overview of ELT courseware in Greek junior high schools and pedagogical alternatives]. Paper presented at the *14th 'Intercultural Education, Immigration, Conflict Management and Pedagogy for Democracy'* international conference, Volos, Greece. 2011.

Kostoulas, Achilleas. "A Greek tragedy: Understanding and challenging 'the Known' from a complexity perspective". In: Rivers, Damian (ed.) *Resistance*

Kostoulas, Achilleas. *A complex systems perspective on English Language Teaching: A case study of a language school in Greece*. Unpublished PhD thesis. The University of Manchester, UK. 2015a.

Kostoulas, Achilleas. "Teaching English to Young Learners in Greece: A critical look". Paper presented at the *49th IATEFL annual convention*, Birmignham, UK. 2015b.

Kostoulas, Achilleas / Mercer, Sarah. "Fifteen years of research on self & identity in System". *System*, 60, 2016, pp. 128–134.

Kostoulas, Achilleas / Stelma, Juup. "Intentionality and complex systems theory: A new direction for language learning psychology". In: Gkonou, Christina / Tatzl, Dietmar / Mercer, Sarah (Eds.), *New Directions in Language Learning Psychology*. Cham: Springer. 2016, pp. 7–22.

Kostoulas, Achilleas / Stelma, Juup. "Understanding curriculum change in an ELT school in Greece". *ELT Journal*, 71(3), 2017, pp. 354–363.

Kostoulas, Achilleas / Stelma, Juup / Mercer, Sarah / Cameron, Lynne / Dawson, Susan. Complex Systems Theory as a shared discourse space for TESOL. *TESOL Journal*. 2017. Advance Access. doi: 10.1002/tesj.317

Kramsch, Claire. *Context and culture in language teaching*. Oxford: Oxford University Press. 1993.

Kramsch, Claire. "The cultural component of language teaching". *Language, Culture & Curriculum*, 8(2), 1995, pp. 83–92.

Kramsch, Claire. "Ecological perspectives on foreign language education". *Language Teaching*, 41(3), 2008, pp. 389–408.

Kuiper, Koenraad. "Studying New Zealand English". *English Today*, 19(3), 2003, pp. 31–34.

Kumaravadivelu, Bala. "The postmethod condition: (E)merging strategies for second/foreign language teaching". *TESOL Quarterly, 28*(1), pp. 27–48.

Kumaravadivelu, Bala. "Toward a postmethod pedagogy". *TESOL Quarterly, 35*(4), 2001, pp. 537–560.

Kumaravadivelu, Bala. *Beyond methods: Macrostrategies for language teaching*. New Haven: Yale University Press. 2003.

Kumaravadivelu, Bala. "Dangerous liaison: Globalisation, empire and TESOL". In: Edge, Julian, (Ed.), *(Re)Locating TESOL in an age of empire* (). Basingstoke: Palgrave Macmillan. 2006a, pp. 1–26

Kumaravadivelu, Bala. "TESOL methods: Changing tracks, challenging trends". *TESOL Quarterly*, 40(1), 2006b, pp. 59–81.

Kumaravadivelu, Bala. *Understanding language teaching: From method to postmethod.* Mahwah, NJ: Lawrence Erlbaum Associates. 2006c

Labov, William. *The social stratification of English in New York city.* Washington, D.C.: Centre for Applied Linguistics. 1966.

Labov, William. *Sociolinguistic patterns.* Philadelphia, PA: University of Pennsylvania Press. 1973.

Lakoff, George. "Cognitive models and prototype theory". In: Margolis, Eric / Laurence, Stephen (Eds.), *Concepts: Core readings.* Cambridge, MA: M.I.T. Press. 1999, pp. 391–421.

Laplace, Pierre-Simon, Marquis de. *Essai philosophiques sur les probabilités [Philosophical Essay on Probabilities].* Paris: Courcier. 1814. Retrieved on 9 August 2014 from: https://archive.org/details/essaiphilosophiq00lapluoft

Larsen-Freeman, Dianne. "Chaos / complexity science and second language acquisition". *Applied Linguistics,* 19(2), 1997, pp. 141–165.

Larsen-Freeman, Dianne. "Grammar". In: Carter, Ronald / David Nunan (Eds.), *The Cambridge guide to teaching English to speakers of other languages.* Cambridge: Cambridge University Press. 2001, pp. 34–41.

Larsen-Freeman, Diane. "Language acquisition and language use from a chaos / complexity theory perspective". In: Kramsch, Claire (Ed.), *Language acquisition and language socialisation.* London: Continuum. 2002.

Larsen-Freeman, Diane. *Teaching language: From grammar to grammaring.* Boston, MA: Thomson/Heinle. 2003.

Larsen-Freeman, Diane. "A complexity theory approach to second language development / acquisition". In: Atkinson, Dwight (Ed.), *Alternative approaches to second language acquisition.* London: Routledge. 2011, pp. 49–72

Larsen-Freeman, Diane. "Classroom-oriented research from a complex systems perspective". *Studies in Second Language Learning and Teaching,* 6(3), 2016, pp. 377–393.

Larsen-Freeman, Diane. / Cameron, Lynne. *Complex systems and applied linguistics.* Oxford: Oxford University Press. 2008.

Lee, Woo-Joo. *Views and experiences of English language education for young learners in South Korea.* Unpublished PhD thesis, University of Manchester, United Kingdom. 2010.

Lee, Winnie Yuk-chun. Authenticity revisited: Text authenticity and learner authenticity. *ELT Journal,* 49(4), 1995, pp. 323–328.

Lenneberg, Eric H. *The biological foundations of language.* New York: Wiley. 1967.

Leung, Constant "Convivial communication: Recontextualizing communicative competence". *International Journal of Applied Linguistics*, 15(2), 2005, pp. 119-144.

Leung, Ching Yin / Andrews, Stephen. "The mediating role of textbooks in high-stakes assessment reform". *ELT Journal*, 66(3), pp. 356-365.

Lewis, Michael. *The lexical approach*. Hove: Language Teaching Publications. 1993.

Li, Defeng. "It's always more difficult that you planned: Teachers' perceived difficulties in introducing the communicative approach in South Korea". TESOL Quarterly, 1998, pp. 677-703.

Lin, Angel / Martin, Peter W. *Decolonisation, globalisation: Language in education policy and practice*. Clevedon: Multilingual Matters. 2005.

Littlejohn, Andrew L. *Why are ELT materials the way they are?* Unpublished PhD thesis, Lancaster University. 1992.

Littlejohn, Andrew L. "The analysis of language teaching materials: Inside the Trojan Horse". In: Tomlinson, Brian R. (Ed.), *Materials development in language teaching*. Cambridge: Camdridge University Press. 1998.

Littlewood, William. *Communicative language teaching: An introduction*. Cambridge: Cambridge University Press. 1981.

Liu, Dilin. "Comments on B. Kumaravadivelu's 'The postmethod condition: (e)merging strategies for second/foreign language teaching "alternative to" or "addition to" method?'" *TESOL Quarterly*, 29(1), 1995, pp. 174-177.

Long, Michael H. "The role of instruction in Second Language Acquisition". In: Hyltenstam, Kenneth / Pienemann, Manfred (Eds.), *Modeling and assessing second language acquisition*. Clevedon: Multilingual Matters. 1985, pp. 77-100.

Long, Michael H. / Sato, Charlene J. "Foreigner talk discourse: Forms and functions of teachers' questions". In: Seliger, Herbert, W / Long, Michael H. (Eds.), *Classroom-oriented research on Second Language Acquisition*. Rowley, MA: Newbury House. 1983, pp. 268-285.

Lorenz, Edward. *Does the flap of a butterfly's wings in Brazil set off a tornado in Texas?* Paper presented at the 139th annual meeting of the American Association for the Advancement of Science, Washington, D.C.. 1972.

Lortie, Dan C. *Schoolteacher: A sociological study*. Chicago: University of Chicago Press. 2002[2].

Lyneis, James M. / Cooper, Kenneth G. / Els, Sharon A. "Strategic management of complex projects: A case study using system dynamics". *System Dynamics Review*, 17(3), 2001, pp. 237-260.

Lyotard, Jean-François. *The postmodern condition: A report on knowledge*. Manchester: Manchester University Press. 1984.

Mason, Mark. *Complexity theory and the philosophy of education*. Oxford: Wiley-Blackwell. 2008.

Matsangouras, Ilias G. *Στρατηγικές διδασκαλίας [Teaching strategies]*. Athens: Gutenberg. 1988.

Matsangouras, Ilias G. «Πρακτική, επιστημονίζουσα και επιστημονική διδακτική» [Practical, science-like and scientific didactics]. In: Matsangouras, Ilias (Ed.), *Η εξέλιξη της διδακτικής: επιστημονική θεώρηση [The evolution of didactics: a scientific perspective]*. Athens: Gutenberg. 1995, pp. 33–76.

Matsuda, Aya. "Incorporating World Englishes in teaching English as an international language". *TESOL Quarterly*, 37(3), 2003, pp. 719–729.

Mattheoudakis, Marina. "Tracking changes in pre-service EFL teacher beliefs in Greece: A longitudinal study". *Teaching & Teacher Education*, 23(8), 2007, pp. 1272–1288.

Mattheoudakis, Marina / Alexiou, Thomais. "Early foreign language instruction in Greece: socioeconomic factors and their effect on young learners' language development". In: Nikolov, Marianne (Ed.), *The age factor and early language learning* (). Berlin: De Gruyter Mouton. 2009, pp. 227–252.

Mauranen, Anna. "The corpus of English as lingua franca in academic settings". *TESOL Quarterly,* 37(3), 2003, pp. 513–527.

Mauranen, Anna. "A rich domain of ELF: The ELFA corpus of academic discourse". *Nordic Journal of English Studies*, 5(2), 2006, pp. 145–159.

Mauranen, Anna. *Exploring ELF: Academic English shaped by non-native speakers*. Cambridge: Cambridge University Press. 2012.

Maxwell, Joseph A. "Understanding and validity in qualitative research". In: Huberman, Michael A. / Miles, Mathew B. (Eds.), *The qualitative researcher's companion*. Thousand Oaks, CA: SAGE. 2002, pp. 37–64.

McArthur, Tom. *The English languages*. Cambridge: Cambridge University Press. 1998.

McDonough, Jo / Shaw, Christopher. *Materials and methods in ELT: A teacher's guide*. Oxford: Blackwell. 1993.

McKay, Sandra L. "EIL curriculum development". In Rubdy Rani, / Saraceni, Mario (Eds.), *English in the world: Global roles, global rules*. London: Continuum. 2006, pp. 130–150.

Meara, Paul. "Towards a new approach to modelling vocabulary acquisition". In: Schmidt, Norbert / McCarthy, Michael (Eds.), *Vocabulary: description, acquisition and pedagogy*. Cambridge: Cambridge University Press. 1997, pp. 109–121.

Meara, Paul. "Modelling vocabulary loss". *Applied Linguistics,* 25(2), 2004, pp. 137–155.

Meara, Paul. "Emergent properties of multilingual lexicons". *Applied Linguistics,* 27(4), 2006, pp. 620–644.

Melka, Francince. "Receptive vs. productive aspects of vocabulary". In: Schmidt, Norbert / McCarthy, Michael (Eds.), *Vocabulary: Description, acquisition and pedagogy* (). Oxford: Oxford University Press. 1997, pp. 84–102.

Mellow, J. Dean. "The emergence of second language syntax: A case study of the acquisition of relative clauses". *Applied Linguistics,* 27(4), 2006, pp. 645–670.

Mellow, J. Dean / Reeder, Kenneth / Forster, Elisabeth. "Using time-series research designs to investigate the effects of instruction on SLA". *Studies in Second Language Acquisition,* 18(3), 1996, pp. 325–350.

Mercer, Sarah. "Language learner self-concept: Complexity, continuity and change". *System,* 39(3), 2011a, pp. 335–346.

Mercer, Sarah. *Towards an understanding of language learner self-concept.* Cham: Springer. 2011b.

Mercer, Sarah. "Understanding learner agency as a complex dynamic system". *System,* 39(4), 2011c, pp. 427–436.

Mercer, Sarah. "Self-concept: Situating the self". In: Mercer, Sarah / Ryan, Stephen / Williams, Marion (Eds.), *Psychology for language learning: Insights from research, theory and practice.* Houndmills: Palgrave Macmillan UK. 2012, pp. 10–25.

Mercer, Sarah. "Towards a complexity-informed pedagogy for language learning". *Revista Brasileira de Linguística Aplicada,* 13, 2013, pp. 376–398. Retrieved on 9.8.2014 from: http://www.scielo.br/pdf/rbla/v13n2/03.pdf

Mercer, Sarah. "Complexity theories and language teaching: Bridging the gap between theory and practice". Paper presented at the 10th Annual Conference of the British Association of Applied Linguistics Language Learning and Teaching Special Interest Group 'Recognising complexity in language learning and teaching', Leeds. 2014.

Mercer, Sarah. "The contexts within me: L2 self as a complex dynamic system". In: King, Jim (Ed.), *The dynamic interplay between context and the language learner.* Basingstoke: Palgrave Macmillan UK. 2016, pp. 11–28.

Mesthrie, Rajend. "'Death of the mother tongue?' Is English a glottophagic language in South Africa?" *English Today,* 24(2), 2008, pp. 13–19.

Meyler, Michael. "Sri Lankan English: A distinct South Asian variety". *English Today,* 25(4), 2009, pp. 55–60.

Miles, Mathew B. / Huberman, Micheal A. *Qualitative data analysis: An expanded sourcebook.* London: SAGE. 1994².

Miller, John H. / Page, Scott E. *Complex adaptive systems: An introduction to computational models of social life.* Oxford: Princeton University Press. 2007.

Mitchell, Melanie. *Complexity: A guided tour.* New York: Oxford University Press. 2009.

Modiano, Marko. "International English and the global village". *English Today,* 15(2), 1999a, pp. 22–34.

Modiano, Marko. "Standard English(es) and educational practices for the world's lingua franca". *English Today,* 15(4), 1999b, pp. 3–12.

Modiano, Marko. "Inclusive/exclusive? English as a lingua franca in the European Union". *World Englishes,* 28(2), 2009, pp. 208–223.

Mohanan, Karuvannur Puthanveettil. "Emergence of complexity in phonological development". In: Ferguson, Charles Albert / Menn, Lise / Stoel-Gammon, Carol. (Eds.), *Phonological Development: Models, Research, Implications.* Timonium, MD: York Press. 1992, pp. 635–662.

Morin, Edgar. "Restricted complexity, general complexity". Paper presented at the *Colloquium 'Intelligence de la complexite : epistemologie et pragmatique',* Cerisy-La-Salle, France. 2006. Retrieved on 9.8.2014 from: http://www.cogprints.org/5217/1/morin.pdf.

Mufwene, Salikoko S. *The ecology of language evolution.* Cambridge: Cambridge University Press. 2001

Mufwene, Salikoko S. "Colonisation, globalisation, and the future of languages in the twenty-first century". *International Journal of Multicultural Societies,* 4(2), 2002, pp. 162–193.

Murray, Heather. "Swiss English teachers and Euro-English: Attitudes to a non-native variety". *Bulletin Suisse de Linguistique Appliquée,* 77, 2003, pp. 147–165. Retrieved on 9.8.2014 from: http://doc.rero.ch/record/11876/files/bulletin_vals_asla_2003_077.pdf?version=1

National and Kapodistrian University of Athens, Research Board. Ανακοίνωση – Θέσεις Εργασίας [*Announcement – Employment posts*]. Athens: Author. 2010. Retrieved 14.8.2014 from http://www.edulll.gr/wp-content/uploads/2010/10/EKPA_PROSKLHSH_8ESEWN_ERGASIAS_KPG.doc

Nault, Derrick. "Going global: Rethinking culture teaching in ELT contexts". *Language, Culture & Curriculum,* 19(3), 2006. pp. 314–328.

Ngũgĩ, wa Tjiong'o. *Decolonising the mind: The politics of language in African literature*. London: Heinemann/Currey. 1986.

Nicolis, Gregoire. *Introduction to nonlinear science*. Cambridge: Cambridge University Press. 1995.

Nikander, Pirjo. "Working with transcripts and translated data." *Qualitative Research in Psychology* 5(3), 2008, pp. 225–231.

Nikolaidis, Katerina / Mattheoudakis, Marina. "Utopia vs. reality: The effectiveness of in-service training courses for EFL teachers". *European Journal of Teacher Education*, 31(3), 2008, pp. 279–292.

Niyogi, Partha / Berwick, Robert C. "A dynamical systems model for language change". *Complex Systems*, 11(3), 1997, 161–204.

Noutsos, M. *Διδακτικοί στόχοι και το αναλυτικό πρόγραμμα [Teaching objectives and the curriculum]*. Athens – Ioannina: Dodone. 1983.

Nunan, David. *Language teaching methodology*. Harlow: Longman. 1991.

Nunan, David. *Task-based language teaching*. Cambridge: Cambridge University Press. 2004.

O'Regan, John P. "English as a lingua franca: An immanent critique". *Applied Linguistics*, 35(5), 2014, pp. 533–552.

Olk, Harald Martin. "Translating culture – a think-aloud protocol study". *Language Teaching Research*, 6(2), 2002, pp. 121–144.

Onat-Stelma, Zeynep. *Moving from teaching older learners to young learners: Cases of English language teachers in Turkey*. Unpublished PhD Thesis, University of Leeds, United Kingdom. 2005.

Ortega, Lourdes / Iberri-Shea, Gina. "Longitudinal reserach in second language acquisition: recent trends and future directions". *Annual Review of Applied Linguistics*, 25, 2005, pp. 26–45.

Osberg, Deborah / Biesta, Gert. "The emergent curriculum: Navigating a complex course between unguided learning and planned enculturation". *Journal of Curriculum Studies*, 40(3), 2008, pp. 313–328.

Osberg, Deborah / Biesta, Gert. (Eds.). *Complexity theory and the politics of education*. Rotterdam: Sense Publishers. 2010.

Pakir, Anne. "English as a lingua franca: Analyzing research frameworks in international English, World Englishes, and ELF". *World Englishes*, 28(2), 2009, pp. 224–235.

Papadopoulou, Marianna. "The ecology of role play: Intentionality and cultural evolution". *British Educational Research Journal*, 38(4), 2011, pp. 575–592.

Papaefthymiou-Lytra, Sophia. "Foreign language testing and assessment in Greece: An overview and appraisal". *Research Papers in Language Teaching & Testing*, 3(1), 2012, pp. 22–32.

Papageorgiou, G. *Η γλώσσα στο δημοτικό σχολείο [Language in primary schools]*. Athens: Smyrniotakis. 1993.

Papageorgiou, M. *The concept of coherence in the design of EFL programmes with particular reference to General English courses in the private sector.* Unpublished M.Ed. dissertation, Hellenic Open University, Patras, Greece. 2002.

Peacock, Matthew. "The effect of authentic materials on the motivation of EFL learners". *ELT Journal*, 51(2), 1997, pp. 144–156.

Pedagogical Institute. *Διαθεματικό ενιαίο πλαίσιο προγραμμάτων σπουδών* [Interdisciplinary comprehensive curriculum of studies] (Vol. B). Athens: Author. 2003.

Pedagogical Institute. (n.d.). *Οι ξένες γλώσσες στο σχολείο: οδηγός του εκπαιδευτικού ξένων γλωσσών [Foreign languages at school: A foreign language teacher's guide]*. Athens: OEDB.

Peltzer-Karpf, Annemarie. *Selbstorganisationsprozesse in der sprachlichen Ontogenese: Erst- und Fremdsprache(n)*. Tübingen: Narr. 1990.

Peltzer-Karpf, Annemarie. Early foreign language learning: The biological perspective. Background paper for the 'Educational research workshop on the effectiveness of modern language learning and teaching' (Graz, Austria, 5–8 March 1996). Strasbourg: Council for Cultural Cooperation. 1996.

Peltzer-Karpf, Annemarie. "The self-organization of dynamic systems: Modularity under scrutiny". In: Gontier, Natthalie / Bendegem, Jean-Paul / van Aerts, Diedrick, (Eds.), *Evolutionary epistemology, language and culture: A non-adaptationist, systems theoretical approach*. Amsterdam: Springer. 2006, pp. 227–256.

Peltzer-Karpf, Annemarie. "Dynamic systems theory (DST) applied to the evolution of language". In: Aiello G. (Ed.) *Abstracts of the 4th International Nonlinear Science Conference*. 2010.

Peltzer-Karpf, Annemarie. The dynamic matching of neural and cognitive growth cycles. *Nonlinear Dynamics – Psychology and Life Sciences*, 16(1), 2012, pp. 61–78.

Pennycook, Alastair. *The cultural politics of English as an international language*. London: Longman. 1994.

Pennycook, Alastair. *Critical applied linguistics: A critical introduction*. London: Laurence Erlbaum Associates. 2001.

Pennycook, Alastair. "The modern Mission: The language effects of christianity". *Journal of Language, Identity & Education*, 4(2), 2005, pp. 137–155.

Pennycook, Alastair. *Global Englishes and transcultural flows*. London: Routledge. 2007.

Pennycook, Alastair. / Coutand-Marin, Sophie. "Teaching English as a missionary language". *Discourse: Studies in the Cultural Politics of Education*, 24(3), 2003, pp. 337–353.

Phillipson, Robert. *Linguistic imperialism*. Oxford: Oxford University Press. 1992.

Phillipson, Robert. "Voice in global English: Unheard chords in crystal loud and clear". *Applied Linguistics*, 20(2), 1999, pp. 265–276.

Phillipson, Robert. "English or 'no' to English in Scandinavia?" *English Today*, 17(2), 2001, pp. 22–28.

Phillipson, Robert. *English-only Europe? Challenging language policy*. London: Routledge. 2003.

Phillipson, Robert. "English in globalization: Three approaches". *Journal of Language, Identity & Education*, 3(1), 2004, pp. 73–84.

Phillipson, Robert. "English: No longer a foreign language in Europe?" In: Cummins, Jim / Davison Chris (Eds.), *The international handbook of English Language Teaching*. Norwell, MA: Springer. 2007, pp. 123–136.

Phillipson, Robert. "English in globalisation, a lingua franca or a lingua frankensteinia?" *TESOL Quarterly*, 43(2), 2009a, pp. 335–339.

Phillipson, Robert. *Linguistic imperialism continued*. New York: Routledge. 2009b.

Poincaré, Henry. *Science et methode*. Publisher and place of publicaiton not indicated. 1903. Retrieved on 12 February 2014 from: http://www.ac-nancy-metz.fr/enseign/philo/textesph/Scienceetmethode.pdf.

Polat, Brittany / Kim, Youjin. "Dynamics of complexity and accuracy: A longitudinal case study of advanced untutored development". *Applied Linguistics*, 35(2), 2014, pp. 184–207.

Pozoukidis, N. / Babalanidou, Z. «Η διαπολιτισμική διάσταση των νέων βιβλίων αγγλικής γλώσσας για το δημοτικό σχολείο: μια ανάλυση περιεχομένου». [The intercultural dimension of the new English language texbooks for primary schools: A content analysis] *Proceedings of the 13th international conference on Intercultural Education, Immigration and Conflict Management*. Alexandroupolis, Greece. 2010, pp. 340–351.

Prabhu, N. S. *Second language pedagogy*. Oxford: Oxford University Press. 1987.

Prabhu, N. S. "The dynamics of the language lesson". *TESOL Quarterly*, 26(2), 1992, pp. 225–241.

Prigogine, Ilya. / Stengers, Isabelle. *Order out of chaos: Man's new dialogue with nature*. Boulder, CO.: New Science Library. 1984.

Prodromou, Luke. "Defining the successful 'bilingual speaker' of English". In: Rubdy Rani / Sarakeni, Mario (Eds.), *English in the world: global rules, global roles*. London: Continuum. 2006.

Prodromou, Luke. "Is ELF a variety of English?" *English Today*, 23(2), 2007, pp. 47–53.

Prodromou, Luke. (2008). *English as a lingua franca : A corpus-based analysis*. London: Continuum.

Prodromou, Luke / Mishen, Freda. Materials used in Western Europe. In Tomlinson, Brian R. (Ed.), *English language learning materials: A critical review*. London: Continuum. 2008.

Punch, Keith. *Introduction to social research: Quantitative and qualitative approaches*. London: SAGE. 2005^2.

Purgason, Kitty B. "A clearer picture of the 'Servants of the Lord'". *TESOL Quarterly*, 38(4), 2004, pp. 711–713.

Purgason, Kitty B. "Is English a force for good or bad?" *International Journal of Christianity and ELT*, 1(1), 2014, pp. 60–81.

Qiong, Hu Xiao. "Why China English should stand alongside British, American, and the other 'World Englishes'". *English Today*, 20(2), 2004, pp. 26–33.

Quirk, Randolph. "The English language in a global context". In: Quirk, Randolph / Widdowson, Henry G. (Eds.), *English in the world: teaching and learning of language and literature*. Cambridge: Cambridge University Press. 1985, pp. 1–6.

Quirk, Randolph. "Language varieties and standard language". *English Today*, 21(1), 1990, pp. 3–10.

Rajagopalan, Kanavillil. "The language issue in Brazil: When local knowledge clashes with expert knowledge". In: A. S. Canagarajah (Ed.), *Reclaiming the local in language policy and practice*. Mahwaj, NJ: Lawrence Erlbaum Associates. 2005, pp. 99–122.

Ramanathan, Vaidehi. *The English-vernacular divide: Postcolonial language politics and practice*. Clevedon: Multilingual Matters. 2005.

Ramanathan, Vaidehi / Atkinson, Dwight. "Ethnographic approaches and methods in L2 writing research: A critical guide and review". *Applied Linguistics*, 20(1), 1999, pp. 44–70.

Randall, Mick / Samimi, Mohammad Amir. "The status of English in Dubai". *English Today*, 26(1), 2010, pp. 43–50.

Rasch, William / Wolfe, Carry (Eds.). *Observing complexity*. Minneapolis, MN: University of Minnesota Press. 2000.

Reali, Florencia / Christiansen, Morten H. "Sequential learning and the interaction between biological and linguistic adaptation in language evolution". *Interaction Studies*, 10(1), 2009, pp. 5–30.

Reed, Michael / Harvey, David L. "The new science and the old: Complexity and realism in the social sciences". *Journal for the Theory of Social Behaviour*, 22, 1992, pp. 356–379.

Revanoglou, A. M. (n.d.). *Διδακτική Αρχαίων Ελληνικών – Εισαγωγική επιμόρφωση [Teaching Ancient Greek – Teacher Induction Programme]*. PEK Kozanis. Kozani, Greece. Retrieved on 9.8.2014 from: http://www.scribd.com/doc/44939367/Διδακτική-Αρχαίων-Ελληνικών-Εισαγωγική-Επιμόρφωση

Richards, Jack C. / Rodgers, Theodore S. *Approaches and methods in language teaching*. Cambridge: Cambridge University Press. 2014³.

Richards, Keith. *Opening the staffroom door: Aspects of collaborative interaction in a small language school*. Unpublished PhD thesis, University of Aston, Birmingham. 1996.

Rosch, Eleanor. "Principles of categorisation". In: Margolis, Eric / Laurence, Stephen (Eds.), *Concepts: Core readings*. Cambridge, MA: M.I.T. Press. 1999, pp. 189–206

Rosen, Robert. "Some epistemological issues in physics and biology". In: Hiley, Basil / Peat, David F. (Eds.), *Quantum implications: Essays in honour of David Bohm*. London: Routledge. 1987, pp. 314–327.

Ryan, Stephen / Dörnyei, Zoltán. "The long-term evolution of language motivation and the L2 self". In: Berndt, Annette (Ed.), *Fremdsprachen in der Perspektive lebenslangen Lernens*. Frankfurt: Peter Lang. 2013, pp. 89–100

Saleem, Mehvish. *Exploring ESL teachers' psychology from a holistic perspective*. Unpublished doctoral dissertation. University of Graz, Austria. 2018.

Sampson, Richard J. *Complexity in classroom foreign language learning motivation: A practitioner perspective from Japan*. Bristol: Multilingual Matters. 2016.

Saraceni, Mario. "English as a lingua franca: Between form and function". *English Today*, 24(2), 2008, pp. 20–26.

Sawyer, Robert Keith. *Social emergence: Societies as a complex system*. Cambridge: Cambridge University Press. 2005.

Scholfield, P. J. / Gitsaki, C. "What is the advantage of private instruction? The example of English vocabulary learning in Greece". *System*, 24(1), 1996, pp. 117–127.

Scholz, Kyle. "Encouraging free play: Extramural digital game-based language learning as a complex adaptive system". *calico journal,* 34(1), 2017, pp. 39–57.

Schreier, Daniel. "Assessing the status of lesser-known varieties of English". *English Today,* 25(1), 2009, pp. 19–24.

Scrivener, Jim. "ARC: a descriptive model for classroom work on language". In: Willis Dave / Willis, Jane R. (Eds.), *Challenge and change in language teaching.* Oxford: Macmillan Heinemann. 1996.

Seargeant, Phillop. "'More English than England itself': The simulation of authenticity in foreign language practice in Japan". *International Journal of Applied Linguistics,* 15(3), 2005, pp. 326–345.

Seidlhofer, Barbara. "Closing a conceptual gap: The case for the description of English as a lingua franca". *International Journal of Applied Linguistics,* 11(2), 2001, pp. 133–158.

Seidlhofer, Barbara. "Habeas corpus and divide et empera". In: Spellman Miller, Kristyan / Thompson Paul, (Eds.), *Unity and diversity in language use.* London: British Association of Applied Linguistics and Continuum. 2002, pp. 198–220.

Seidlhofer, Barbara. "Research perspectives on teaching English as a lingua franca". *Annual Review of Applied Linguistics,* 24(1), 2004, pp. 209–239.

Seidlhofer, Barbara. "English as a lingua franca in the expanding circle: What it isn't". In: Rubdy Rani / Saraceni, Mario (Eds.), *English in the world: Global rules, global roles.* London: Continuum. 2006, pp. 40–50.

Seidlhofer, Barbara. "Common ground and different realities: World Englishes and English as a Lingua Franca". *World Englishes,* 28(2), 2009, pp. 236–245.

Seidlhofer, Barbara. *Understanding English as a lingua franca.* Oxford: Oxford University Press. 2011.

Seow, Anthony. "The writing process and process writing". In: Richards, Jack C. / Renandya, Willy A. (Eds.), *Methodology of Language Teaching.* Cambridge: Cambridge University Press. 2002, pp. 315–320.

Sewell, Andrew. "English as a lingua franca: Ontology and ideology". *ELT Journal,* 67(1), 2013, pp. 3–10.

Shin, Jeeyoung / Eslami, Zohreh R. / Chen, Wen-Chun. "Presentation of local and international culture in current international English-language teaching textbooks". *Language, Culture & Curriculum,* 24(3), 2011, 253–268.

Shulman, Lee S. (1986). "Those who understand: Knowledge growth in teaching". *Educational Researcher,* 15(2), 1986, pp. 4–14.

Sifakis, Nikos. "Teaching EIL – Teaching international or intercultural English? What teachers should know". *System,* 32(2), 2004, pp. 237–250.

Sifakis, Nikos. "Challenges in teaching ELF in the periphery: The Greek context". *ELT Journal,* 63(3), 2009, pp. 230–237.

Sifakis, Nikos / Fay, Richard. "Integrating an ELF pedagogy in a changing world". In: Archibald, Alasdair / Cogo, Alicia / Jenkins, Jennifer (Eds.), *Latest trends in ELF research.* Newcastle-Upon-Tyne: Cambridge Scholars Publishing. 2011, pp. 285–298.

Sifakis, Nikos / Lytra, Vally / Fay, Richard. "English as a lingua franca in an increasingly post-EFL era: The case of English in the Greek state education curriculum". Paper presented at the 3rd *international conference on English as a Lingua Franca,* Vienna. 2010.

Sifakis, Nikos / Sougari, Areti-Maria. "Pronunciation issues and EIL pedagogy in the periphery: A survey of Greek state school teachers' beliefs". *TESOL Quarterly,* 39(3), 2005, pp. 467–488.

Sifakis, Nikos / Sougari, Areti-Maria. "Rethinking the role of teachers' beliefs about their function: a survey of Greek teachers' inner thoughts". Paper presented at the 14th *international conference of the Greek Association of Applied Linguistics,* Thessaloniki. 2007. Retrieved on 9.8. 2014 from: http://www.enl.auth.gr/gala/14th/Papers/English%20papers/Sifakis&Sougari.pdf.

Silver, Christina. "Participatory approaches to social research". In: Gilbert, Nigel G. (Ed.), *Researching social life.* Thousand Oaks, CA: SAGE. 2008^3, pp. 101–124

Silverman, David. *Doing qualitative research: A practical handbook.* London: SAGE. 2005^2.

Skutnabb-Kangas, Tove. *Linguistic genocide in education – or worldwide diversity and human rights.* London: Lawrence Erlbaum Associates. 2000.

Smith, David I. "On viewing learners as spiritual beings: Implications for language educators". *Journal of Christianity and Foreign Languages,* 9, 2008, pp. 34–48.

Soulioti, Evangelia. *Η διδασκαλία της αγγλικής γλώσσας στο ελληνικό κράτος την περίοδο 1913–64: Η περίπτωση του νομού Ιωαννίνων. Μεθοδολογικές, διδακτικές, ιστορικές προεκτάσεις και συγκριτική θεώρηση με σύγχρονες προσεγγίσεις [English langage teaching in Greece from 1913 to 1967: The case of (the) Ioannina prefecture. Methodological, teaching and historical perspectives and comparison with current approaches].* Unpublished PhD thesis, University of Ioannina, Greece. 2007.

Sowden, Colin. "ELF on a mushroom: The overnight growth in English as a lingua franca". *ELT Journal,* 66(1), 2012a, pp. 89–96.

Sowden, Colin. A reply to Alessia Cogo. *ELT Journal,* 66(1), 2012b, pp. 106–107.

Spoelman, Marianne / Verspoor, Marjolin. "Dynamic patterns in development of accuracy and complexity: A longitudinal case study in the acquisition of Finnish". *Applied Linguistics*, 31(4), 2010, pp. 532–553.

Stelma, Juup. *Visualising the dynamics of learner interaction: Cases from a Norwegian language classroom*. Unpublished PhD thesis, University of Leeds, United Kingdom. 2003.

Stelma, Juup. "An ecological model of developing researcher competence: The case of software technology in doctoral research". *Instructional Science*, 39(3), 2011, pp. 367–385.

Stelma, Juup. "Developing intentionality and L2 classroom task-engagement". *Classroom Discourse*, 5(2), 2014, pp. 119–137

Stelma, Juup / Fay, Richard. "Intentionality and developing researcher competence on a UK master's course: An ecological perspective on research education". *Studies in Higher Education*, 39(4), 2014, pp. 517–533.

Stelma, Juup. / Onat-Stelma, Zeynep / Lee, Woo-Joo / Kostoulas, Achilleas. "Intentional dynamics in TESOL: An ecological perspective". *Teacher's College Columbia Working Papers in TESOL and Applied Linguistics*, 15(1), 2015, pp. 14–32.

Stern, Howard H. *Fundamental concepts of language teaching*. Oxford: Oxford University Press. 1983.

Strauss, Anselm L. / Corbin, Janet M. "Grounded theory methodology: an overview". In: Denzin, Norman, K. / Lincoln, Yvonne, S. (Eds.), *Handbook of qualitative research*. Thousand Oaks, CA: SAGE. 1994, pp. 262–272.

Strevens, Peter. "English as an international language: Directions in the 1990s". In: Braj B. Kachru (Ed.), *The other tongue: English accross cultures*. Urbana, IL.: University of Illinois Press. 1992^2, pp. 27–47

Svalberg, Agneta / Askham, Jim. "A dynamic perspective on student language teachers' different learning pathways in a collaborative context". In: King, Jim (Ed.). *The dynamic interplay between context and the language learner*. Basingstoke: Palgrave-Macmillan. 2016, pp. 172–193.

Swain, Merrill. "Communicative competence: Some roles of comprehensible input and comprehensible output in development". In: Gass, Susan / Madden, Carolyn (Eds.), *Input in Second Language Acquisition*. Rowley, MA: Newbury House. 1985, pp. 235–256.

Thelen, Esther / Smith, Linga B. *A dynamic systems approach to the development of cognition and action*. Cambridge, MA: M.I.T. Press. 1994.

Thom, René. *Stabilité structurelle et morphogénèse; essai d'une théorie générale des modèles* [Structural stablity and morphogenesis: An essay on a general theory of modelling]. Reading, MA.: W. A. Benjamin. 1972.

Thom, René. *Mathematical models of morphogenesis*. Chichester: Ellis Horwood. 1983.

Timmis, Ivor. "Native-speaker norms and International English: A classroom view". *ELT Journal*, 56(3), 2002, pp. 240–249.

Tomasello, Michael. *Constructing a language*. Cambridge, MA: Harvard University Press. 2003.

Tribble, Chris. *Writing*. Oxford: Oxford University Press. 1996.

Trudgill, Peter. "Standard language: What it isn't". In: Bex, Tony / Watts, Richard J. (Eds.), *Standard English: The widening debate*. London: Routledge. 2002, pp. 117–128

Tsolakis, C. *Νεοελληνική γραμματική της Ε' και Στ' Δημοτικού [Modern Greek Grammar for Years 5 and 6]*. Athens: OEDB. 1978.

Tudor, Ian. *The dynamics of the language classroom*. Cambridge: Cambridge University Press. 2001

Tudor, Ian. "Learning to live with complexity: Towards an ecological perspective on language teaching". *System,* 31(1), 2003, pp. 1–12.

Turner, F. "Foreword". In: Eve, Raymond A. / Horsfall Sara / Lee, Mary E. (Eds.), *Chaos, Complexity, and Sociology: Myths, Models, and Theories*. Thousand Oaks, CA: SAGE. 1997.

University of Athens Research Centre for English Language. Introducing the KPG speaking test: New oral examiner information pack. Athens: Author. 2012.

University of Athens Research Centre for English Language. Αγγλικά από τόσο νωρίς; [English from such an early age?] n.d. Retrieved on 28.2.2014 from http://rcel.enl.uoa.gr/peap/parents-corner/agglika-apo-toso-noris.

University of Cambridge Local Examinations Syndicate. *First Certificate in English: Handbook for teachers for examinations from December 2008*. Cambridge: Author. 2007.

Valsiner, Jaan. *The guided mind*. Cambridge, MA: Harvard University Press. 1998.

van Lier, Leo. *The classroom and the language learner: Ethnography and second-language classroom research*. London: Longman. 1988.

van Lier, Leo. *The ecology and semiotics of language learning: A sociocultural perspective*. New York, NY: Springer. 2004a.

van Lier, Leo. The semiotics and ecology of language learning: Perception, voice, identity and democracy. *Utbildning & Demokrati,* 13(3), 2004b, pp. 79–103.

Varghese, Manka M. / Johnston, Bill (2007). "Evangelical Christians and English Language Teaching". *TESOL Quarterly,* 41(1), 5–31.

von Bertalanffy, L. An outline for general systems theory. *British Journal for the Philosophy of Science,* 12, 1950. 134–165.

Vougioukas, A. *Το γλωσσικό μάθημα στην πρώτη βαθμίδα της νεοελληνικής εκπαίδευσης [The language lesson in the first stage of modern Greek education].* Thessaloniki: Institute of Modern Greek Studies (M. Triantafillides Foundation). 1994.

Wallace, Michael J. *Action research for language teachers.* Cambridge: Cambridge University Press. 1998.

Widdowson, Henry G. "The teaching of English as communication". *ELT Journal,* 27(1), 1972. pp., 15–19.

Widdowson, Henry G. *Explorations in applied linguistics.* Oxford: Oxford University Press. 1979.

Widdowson, Henry G. "EIL, ESL, EFL: Global issues and local interests". *World Englishes,* 16(1), 1997, pp. 135–146.

Widdowson, Henry G. *Defining issues in English language teaching.* Oxford: Oxford University Press. 2003.

Widdowson, Henry G. "A perspective on recent trends". In: A. P. R. Howatt (Ed.), *A history of English language teaching.* Oxford: Oxford University Press. 2004[2], pp. 353–372

Wilkins, David Arthur. *Notional syllabuses: A taxonomy and its relevance to foreign language curriculum development.* London: Oxford University Press. 1976.

Williams, Ann. "Non-standard English and education". In: Britain, David. (Ed.), *Language in the British Isles.* Cambridge: Cambridge University Press. 2007[2], pp. 401–416.

Willis, Dave / Willis, Jane R. *Doing task-based teaching.* Oxford: Oxford University Press. 2007.

Willis, Jane R. *A framework for task-based learning.* Harlow: Longman. 1996.

Wong, Mary Shepard / Canagarajah, Suresh A. (Eds.). *Christian and critical English language educators in dialogue: Pedagogical and ethical dilemmas.* New York: Routledge. 2009.

Wong, Mary Shepard / Kristjansson, Carolyn, / Dörnyei, Zoltán. (Eds.). *Christian faith and English language teaching and learning: research on the interrelationship of religion and ELT.* New York, NY: Routledge. 2013.

Wong, Viola / Kwok, Peony / Choi, Nancy. "The use of authentic materials at tertiary level." *ELT Journal,* 49(4), 1995, pp. 318–322.

Xanthakou, E. K. (2005). *Learning English in a TENOR situation: A critical examination of teaching English in the Greek state Gymnasio based on a small-scale research*. Unpublished M.Ed. dissertation, Hellenic Open University, Patras, Greece.

van Geert, Paul. "A dynamic systems model of cognitive and language growth". *Psychological Review,* 98(1), 1991, pp. 3–53.

Verspoor, Marjolijn / Lowie, Wander / Van Dijk, Marijn. "Variability in second language development from a dynamic systems perspective". *The Modern Language Journal*, 92(2), 2008, pp. 214–231.

Verspoor, Marjolijn / Smiskova, Hana. "Foreign language writing development from a dynamic usage based perspective". In: Manchón, Rosa M. (Ed.), *L2 writing development: Multiple perspectives*. Berlin: De Gruyter. 2012, pp. 17–46.

White, Cynthia. "Language teacher agency". In: Mercer, Sarah. / Kostoulas, Achilleas (Eds.), *Language Teacher Psychology*. Bristol: Multilingual Matters. 2018, pp. 196–210.

Yin, Robert K. *Case study research: Design and methods*. Thousand Oaks, CA: SAGE. 2003³.

Young, Michael / DePalma, Andrew / Garrett, Steven. "Situations, interaction, process and affordances: An ecological psychology perspective". *Instructional Science,* 30(1), 2002, pp. 47–63.

Yu, Kwan-Mei. *Christianity and English language teaching: A study of an English conversation class for mainland Chinese scholars at an English speaking church in Hong Kong*. Unpublished MA Thesis, University of Hong Kong, Hong Kong. 2007. Retrieved on 9.8.2014 from http://hub.hku.hk/bitstream/10722/51820/6/FullText.pdf?accept=1

Yu, Liming. "Communicative language teaching in China: Progress and resistance". *TESOL Quarterly,* 35(1), 2001, pp. 194–198.

Zacharias, Nugrahenny T. "Teachers' beliefs about internationally-published materials: A survey of tertiary English teachers in Indonesia". *RELC Journal,* 36(1), 2005, pp. 23–37.

Zouganeli, Aikaterini. "Age constraints on first versus second language acquisition: evidence for linguistic placticity". *PE@P,* 1(1). 2011a.

Zouganeli, Aikaterini. «Όσο νωρίτερα, τόσο καλύτερα» [The earlier the better]. *PE@P,* 1(1). 2011b.

Zouganeli, Aikaterini. / Giannakopoulou, A. «Ανάπτυξη σχολικού γραμματισμού και εκμάθηση ξένης γλώσσας» [Developing academic literacy and learning a foreign language]. *PE@P,* 1(1). 2011.

Index

A
accountability 138
affordance landscape 23, 91
affordances
  communicative 110
  in grammar activities 96
  in listening activities 106
  in reading activities 105
  in speaking activities 107
  in vocabulary activities 99
  in writing activities 103
  transmissive 109
  variation 109
agency 52, 57, 84, 115
Anglicism 83
Applied Linguistics (Journal) 37
appropriate methodology 78
assessment 168
attractors 20, 23, 32, 33, 89, 153, 186, 205
  definition 46
  definition of 154
  prevalence in the curriculum 158
  types of 154
authenticity 80
autopoesis 200

B
boundaries 42

C
Cambridge ESOL 122
case study 58
Certificate of Proficiency in English 122
Certification intentionality 117
chaos theory 36, 48
co-adaptation 140
cognitions 52
Communicative Teaching Project 76
Communicative Language Teaching 74
Competition intentionality 136
complex systems
  definition 21
  heterogeneity of 41
  historicity 46
  in critical pedagogy 51
  in education 50
  in ELT 50
  in first language acquisition 48
  in linguistics 48
  in Psychology of Language Learning 51
  in Second language acquisition 49
  resilience of 45
  unpredictability of 142
computational modelling 56
constraining structures 5, 86
control parameter 208
credentialism 117
Critical Discourse Analysis 203
critical pedagogy 51, 81, 84
Cultural Awareness intentionality 130
cultural continuity 85
cultural empathy 83

D
diachronic change 111, 190
Discours de la Méthode 38
dynamic ensemble 55
dynamics of intentions 149, 189, 196, 200, 201, 202, 207

249

E
ecological psychology 90
ecological theory 50
ELT in Greece 117
  history of 206
  primary schools 25
  private education 28, 138
  secondary schools 26
  state education 24, 26, 134, 135, 136
  State education 25
emergence 44, 115, 169, 180
English as a Lingua Franca 67
English Today debate 63, 65
Examination of Communicative Competence in English 123
Examination of Communicative Proficiency in English 123
exploratory practice 78

F
First Certificate in English 122
Five Graces paper 54
framing 22, 42

G
globalisation 206
grammar 171
  in learning materials 96
  perceived value of 71
  practice 175
  teaching 71, 171
grammaring 48
grounded theory 211

H
habitus 47

I
integrative premise 39
intentionality 51, 190
  definition 113, 201

derived 200
orientation 116
International Integration intentionality 127

L
Laplace 39
learning materials
  companions 101
  cultural content of 133
  grammar 96
  in Greek ELT 77, 90, 143, 198
  in the language school 92
  listening 106
  reading 105
  role of 89
  speaking 107
  vocabulary 99
  writing 103
linguistic discrimination 125
Linguistic Imperialism 82
listening 106, 157
London Tests of English 123
Lorenz 36

M
Manchester Roundtable on Complexity and TESOL 55
morphogenesis 44, 47
Multicultural Awareness Through English 85

N
nestedness 43, 114

O
oral method 73
Orientalism 83

P

Pedagogical Content Knowledge 144
phase shift 110, 190, 207
Phonology of English as an International Language 67
Poincaré 39
political aspects of ELT
 awareness 81
 neutrality 80
 resistance 84
post-method macro-strategic framework 78
post-method pedagogy 77
predictive premise 39
prosyletisation 83
Protectionism intentionality 143

R

rationalities 42, 201
reading 105, 159, 162, 165
 in learning materials 105
 teaching practices 159
 text types 105
reciprocal determination 44
reductive premise 38
retrodiction 57, 205
Retrodictive Qualitative Modelling 57
Revista Brasileira de Linguistica Aplicada (Journal) 37
routinisation 154

S

Second Language Acquisition 49
self-organisation 37, 44
sensitive dependence 36
speaking 107, 157
Standard Language Ideology 63, 124
State Certificate of Language Proficiency 123
state space 23, 61

T

Task-Based Learning 76
Teaching English to Young Learners 25, 140
three-body problem 39
Three Circles model 65
time-series analyses 57
tissue rejection 78
transmissive pedagogy 71
tripartite model 73

U

University of Michigan examinations 123

V

vocabulary 99, 159, 166, 168

W

World Englishes 64
writing 103
 form-focussed 104
 genre-based 104, 157
 process-based 104, 157, 181

www.ingramcontent.com/pod-product-compliance
Ingram Content Group UK Ltd.
Pitfield, Milton Keynes, MK11 3LW, UK
UKHW040948220426
5322IPUK00028B/31